Hydrology: A Question of Balance

Hydrology:
A Question of Balance

J. V. Sutcliffe

IAHS AISH

IAHS Special Publication 7

IWMI

The International Water Management Institute, Colombo, Sri Lanka, sponsored this publication.

Published by the International Association of Hydrological Sciences 2004

IAHS Special Publication 7
ISBN 1-901502-77-5

British Library Cataloguing-in-Publication Data.
A catalogue record for this book is available from the British Library.

IAHS Publications are available from:
IAHS Press, Centre for Ecology and Hydrology, Wallingford, Oxfordshire OX10 8BB, UK
tel: +44 1491 692442; fax: +44 1491 692448; e-mail: jilly@iahs.demon.co.uk

Printed in the UK by J W Arrowsmith Ltd

CONTENTS

FOREWORD

This is a unique book. Many of the hydrology texts currently available offer accounts of the principles of hydrology, illustrated by way of examples. In this book, examples are the backbone of the presentation, explaining practical problems to be solved, while the methods that were used to find a solution, are secondary.

The book offers the experience of a hydrologist with extraordinary practical expertise, assembled over more than half a century, from nearly all the continents, in areas with different climates, topographies, socio-economic systems, levels of development, and cultures.

While many of Dr Sutcliffe's works have been conducted in Britain, the thread of hydrological studies has taken him to some 50 countries, including: Egypt, Sudan, Jordan, Abu Dhabi, Bahrain, Yemen, Iran, Senegal, Botswana, Kenya, Uganda, India, Sri Lanka, Argentina, New Zealand, Bosnia and Poland. The Nile basin, including the Jonglei Canal and the Nile wetlands, has played a particular role in his studies and he developed a considerable and long-lasting expertise there. Dr Sutcliffe knows well what he is writing about, having gathered first-hand experience by visiting places, and working in the field, in addition to studying the academic literature, consultant reports and data sets. He has built a mental picture of each investigation area and a feel for the dominant processes.

Basic hydrology is indeed a question of balance. This is the most basic issue. Water balance law must be fulfilled—the principle of the conservation of mass is elementary and powerful with no exceptions. Thanks to the fundamental water balance equation we can determine values of a variable which we do not know, based on the (observed or estimated) values of other variables, that occur in the budget equation. Any apparent departure from the balance simply means that we are in error and that not all the terms in the water balance equation have been adequately evaluated. Hence, checking the water balance is useful to evaluate whether our assessments are realistic. For instance, the explanation of a substantial residual in the water balance can be that evaporation has been either under- or overestimated.

Actually, hydrology comprises much more than just the water balance—there are other important general laws, with the conservation of energy and momentum being of ubiquitous validity. However, if we wish to prioritize and focus on a single, most essential principle, which still enables one to draw useful conclusions, then the water balance is definitely what matters most.

When summarizing a hydrological publication, one usually seeks the principal strengths falling into two major classes: novel and innovative methodology (holding potential for application), and/or contribution to the understanding of hydrology of a particular region. The principal asset of Dr Sutcliffe's book is undoubtedly in the latter category. It covers global hydrology—not in the sense of modelling of the global hydrological cycle, but rather in the sense of presentation of case studies worldwide, hence contributing to understanding of water availability conditions in many regions of the world.

Since the case studies presented in the book span the time interval from the 1950s on, the author has focused on what he actually did rather than what might be done (modern equipment and methodology being available). In a way it is a tribute to what can be done without broader applications of the tools of modern technology—Geographic Information Systems, remote sensing, isotopes, etc. Dr Sutcliffe has had to be a pragmatist and address the problem with what was available at the time, and his best knowledge.

At the advent of the PUB (Prediction in Ungauged Basins) initiative, the current flagship IAHS activity, many a younger PUB expert could learn from Dr Sutcliffe's approach to the intriguing issue: What can be done if there are no data but there is a burning practical problem and a decision must be made anyway, regardless of the state of scientific understanding? Stating "I do not know" is not acceptable—one has to make an "optimal", informed decision, despite the circumstances of high uncertainty. Dr Sutcliffe undertakes a "mission impossible", trying to synthesize scarce information, based on his expert feeling and intuition, common sense and the law of water balance. In one figure in the book, he is happy with ten data points; this is all that was available. Yet, it gave him orientation in understanding the process behaviour. Dr Sutcliffe explains how to augment the database, using auxiliary information from different sources, where detective work, based on common sense and logical thinking is needed. Like a detective, he knows where to look for traces—where water may have left them, and how to look for traces, equipped with an analogue of a magnifying glass—an arsenal of robust hydrological practices. Field reconnaissance and air photography allowed him to estimate transpiration and water use in the Middle East. He estimated flows using areas of perennial date palms and other crops irrigated near villages along the stream. Deciphering evidence from the remote past, he placed ancient water-related infrastructures in central India in their hydrological context.

The question is not only what to do in ungauged, or inadequately gauged, basins, but what to do if the available data are suspected to be wrong and need to be corrected. In order to do so, Dr Sutcliffe has sought ancillary information from various non-standard sources, estimating flow and groundwater recharge in an indirect way. Interdisciplinary contacts with meteorologists, geologists, civil engineers, planners, botanists, and archaeologists allowed him to draw robust corollaries from scarce proxy, and uncertain, information. He has been tuned to receiving useful information when talking to apparently lay people, like a teacher in New Zealand. When he and I chatted together about Polish floods in Poznań, John was seeking information with a vigour and concentration which would be envied by many a young hydrologist.

The name of the author is well known to the hydrological community. Web browsers give multiple entries related to his name. Whether you "google" or "yahoo", in a fraction of a second you will have a myriad of references to Dr John Sutcliffe on your screen (be careful; several individuals bearing this name and surname have made themselves known in various areas of activity). He is often cited in academic publications, most frequently (many hundreds of citations) with reference to his paper published jointly with the late Professor Nash in 1970. However, this paper, containing the formula broadly known as the Nash-Sutcliffe criterion of model performance (also called: goodness-of-fit or efficiency criterion), being widely used and cited over decades, summarizes just a small episode in the author's rich career, which included leadership of the team which produced the *Flood Studies Report* for the British Isles.

Dr Sutcliffe makes interesting comments on the decrease of time available for hydrological investigations. This reminds me of the words I heard in 1969, as a freshman student at the Technical University of Warsaw, from a professor of mathematics: "We do not develop mathematics in order to construct machines, but we develop machines in order to have more time to deal with mathematics." This has turned out to be a largely incorrect statement. The technology is there and it improves the efficiency of our work, but time has become more rare and precious. In the past, time constraints were not that severe. The same pertained to data availability in many regions. The severity of water-related problems grows but, counter-intuitively, fewer ground-truth data are being collected now than in the past. Today, satellite watch embraces the whole globe, but understanding of the local detail is missing. Hydrological services receive inadequate funding and shut down monitoring/ hydrometric stations, even those with long continuous time series of observations. If we only had the time and the data—this is a common longing.

Dr Sutcliffe's hydrology, built around the water balance law, has been applied in vast areas of less developed countries, where water-related problems are many and severe, while the human coping capacity and expertise, and the available data, are scarce. This is where the book is likely to find a considerable resonance. This publication fits the IAHS mission—to enhance development of hydrological sciences worldwide, including less developed countries, where having too little, too much, or too polluted water is commonplace. Experts involved in development assistance will find the book of interest. Among potential readers are also experts in the new member countries of the extended European Union, which are likely gradually to turn from being beneficiaries of international development assistance to providers of international aid.

Dr Sutcliffe has been a prolific author; the bibliography contains several dozen works by him, often in co-authorship. Many occur in IAHS publications, spanning the time frame from 1960 to 2004, including several papers published in *Hydrological Sciences Journal* and in the IAHS Red Book series, and *Hydrology of the Nile* (co-authored with Yvonne Parks), a volume in the IAHS Special Publications series. Unlike many consultants, who produce project reports only (an obligatory element of a solicited study) and are not interested in converting their findings into academic publications (which may require far more work, attention, and pain to get through the peer review process), Dr Sutcliffe has done both. In addition to reports to the funding agencies, after each of the many projects, he has published papers in the professional hydrological science literature. This must have been driven by an internal mechanism—as a consultant he has not been subject to the pressure of the academic "publish or perish" principle.

The book will be of interest to several international initiatives, such as the IAHS PUB initiative mentioned above, the HELP (Hydrology for Environment, Life and Policy) initiative of UNESCO, and the Millennium Development Goals, illustrating the importance of freshwater resource assessment, and of estimating different components of the water balance (both long-term averages and variability).

A common Polish greeting, conveyed in a song used when celebrating someone's nameday, birthday, etc. is a wish for him, to live a hundred years in good health and prosperity. Celebrating the birth of the book, let me echo this Polish wish to Dr Sutcliffe, after having added my hope that he will continue to share his expertise with younger generations, and to provide his valuable input to IAHS activities. I do hope that this book is widely read and used among the global hydrological community, for it will bring benefits by improving our understanding of the water cycle.

Zbigniew W. Kundzewicz
IAHS Editor
September 2004

PREFACE

This book is intended to bring hydrological analysis to life by means of examples in which the author has been involved, either directly in solving problems or in terms of review. It is not intended to take the place of standard textbooks, and indeed this book presupposes a reasonable knowledge of basic hydrology. There are many excellent textbooks and papers in which methods of analysis are presented: examples which spring to mind are the World Meteorological Organization's *Guide to Hydrological Practices* (fifth edition, 1994), Linsley, Kohler & Paulhus: *Hydrology for Engineers* (1958), Chow: *Handbook of Applied Hydrology* (1964), Chow, Maidment & Mays: *Applied Hydrology* (1988), Maidment: *Handbook of Hydrology* (1993), Wilson: *Engineering Hydrology* (fourth edition, 1990), Shaw: *Hydrology in Practice* (third edition, 1994), Ward & Robinson: *Principles of Hydrology* (fourth edition, 2000). More specialised topics are covered in such publications as Beven: *Rainfall–Runoff Modelling: The Primer* (2001), Abbott & Refsgaard: *Distributed Hydrological Modelling* (1996), Vieux: *Distributed Hydrological Modeling Using GIS* (2001), Bras & Rodriguez-Iturbe: *Random Functions and Hydrology* (1994), Rushton: *Groundwater Hydrology: Conceptual and Computational Models* (2003), Eagleson: *Ecohydrology*, (2002), Shahin: *Hydrology and Water Resources of Africa* (2002), and Newson: *Land, Water and Development* (1992). Textbooks usually concentrate on accepted methods of analysis, which are followed by accounts of examples. The intention here is to reverse this procedure, and to describe the problem and then the approach chosen in the particular example. There are usually several approaches which could be used to solve a problem, and to some extent the method of analysis depends on the context of the problem. Different methods are likely to be appropriate at different stages of a project; the detailed analysis required for a final design is not needed for a preliminary study to determine whether a project is likely to be viable in hydrological and economic terms.

It is a fact of modern financial pressure that the time available for hydrological investigation and analysis has decreased greatly over the years. It could be argued that the greater availability of records and the power of modern computing facilities compensate for the shorter time scale of investigations; however, the time available for a study affects the type and depth of analysis which can be carried out. The most appropriate method may have changed over the years as new techniques have been developed. However, the solution which seemed appropriate when the work was carried out is described here, with some comment on how a similar problem would now be approached.

It is the belief of the author that the application of a water balance is essential to the understanding of most problems in applied hydrology. The position of the project area on the scale between the humid and arid ends of the climate spectrum will determine the relative importance of the different components and will affect the complexity of the problem. At the outset of his career, he was confronted with situations as dissimilar as the swamps of the upper Nile, the water balance of the semi-arid Alborz mountains in Iran, the seasonal balance of the central highlands of Kenya, and the humid regime of the forested and snow-fed volcanic uplands of the North Island of New Zealand. The context of each study varied. The first was an early environmental study of the Jonglei Canal, for which a team of some ten professionals was able to devote four years of research. The others included a study of runoff and groundwater recharge in Iran, a comparison between upstream and downstream water use in Kenya, and the investigation of hydroelectric development in New Zealand. The common factor was the application of a water balance approach, which provided estimates of the variables of direct interest in each case.

A career spanning over 50 years has included an alternation of rapid applied investigations and longer periods of research into problems at greater depth. The range of climates covered has varied from the arid or semiarid conditions of Botswana, Bahrain and Yemen, through the seasonal conditions of central India and the countries of West Africa, to the more humid climates of Sri Lanka, the mountains of Bosnia and the Patagonian foothills of the Andes.

Although the early studies would have been made easier with the help of personal computers, as the more recent investigations have been facilitated, the fundamental problems of understanding and quantifying the physical processes have not changed. Modern technology may enable more complex means of estimating the components of the water balance to be used, but these are not always appropriate to developing countries where data are limited. Many of the examples come into this category, and it is unfortunately true that the quantity of hydrological records is decreasing in many countries. It is hoped that accounts of the problems encountered in areas as diverse as the Nile (1950s), Iran (1960s), India (1970s), West Africa (1980s) and Sarajevo and central Europe (1990s) will provide an illustrated complement to the more formal examples in the literature.

In the studies leading to this book, acknowledgements are due to a large number of colleagues over the years, especially those with whom the author has worked at the Institute of Hydrology (now CEH), Wallingford, and at Sir Alexander Gibb & Partners (now Jacobs), Reading, UK. This collaboration is illustrated by the number of co-authors cited in the references. It has always been the author's policy to publish the results of a project where it seemed interesting or useful to others. Many references are made to these previous publications as more details are often given there. Thanks are also due to hydrologists and meteorologists in many countries.

The author is particularly grateful to Yvonne Parks for her encouragement and helpful, if sometimes caustic, comments on versions of the manuscript, and to Steve Dunthorne and others for their comments. He is also grateful to Cate Gardner and Sonja Folwell for their help in producing the final version.

J. V. Sutcliffe
Goring on Thames, Oxfordshire
September 2004

ACKNOWLEDGEMENTS

The author and publishers are grateful to the following for permission to reproduce copyright material, and to the respective authors:

The British Hydrological Society (BHS) for Figs 4.2, 8.7 and 8.8 (all from Parks & Sutcliffe, 1987); Cambridge University Press for Fig. 11.9 (Howell *et al.*, 1988); the Chartered Institution of Water and Environmental Management (CIWEM) for Fig. 6.7 (Penman, 1950); Elsevier for Fig. 6.5 (Smith & Piper, 1978), Fig. 8.25 (Mandeville *et al.*, 1970), and Figs 10.9–10.11 (Sutcliffe, 1987); John Wiley and Sons Ltd for Fig. 3.1 (Lazenby & Sutcliffe, 1994); the Royal Geographical Society for Fig. 6.6 (Beaumont, 1968); the Society for South Asian Studies for Plate 8.6 (Shaw & Sutcliffe, 2001); Thomas Telford Publishing for Figs 8.2 and 8.3 (Piper *et al.*, 1994); UNESCO Division of Water Sciences for Figs 10.6 and 10.7 (Sutcliffe & Farquharson, 1998); the World Conservation Union (IUCN) for Figs 8.31 and 11.8 (Sutcliffe & Parks, 1996).

Many other figures are reproduced from papers in *Hydrological Sciences Journal* and other IAHS Press publications, and we are grateful to their authors too.

We are grateful to Martin G. Hodnett for providing the photograph in Plate 5.1. Julia Shaw took the photograph in Plate 8.6. Other photographs were taken by the author.

1 NEED FOR HYDROLOGICAL INFORMATION

Although hydrology as a science includes all aspects of the water cycle, it is generally applied in the development of water resources or flood control. The design of a water resources project, or the assessment of the water resources of an extended area, requires a network of hydrological records that is both reliable in quality and reasonably long in duration. In this chapter the types of hydrological data needed for different types of surface water resources project are discussed. In general terms measurements should be adequate to estimate the different components of the water balance in the project area, including both long-term averages and variations over time.

THE WATER BALANCE

One of the main scientific bases of hydrology is the knowledge that the inflows to any area over any period of time must equal the outflows plus the change in water storage. In general terms the precipitation, in the form of rainfall or snowfall, must equal evaporation and river or groundwater outflow, plus changes in soil moisture, groundwater and channel storage. Thus measurements of all the components of the balance can provide a check on the accuracy of the estimates, or alternatively one of the components of the balance can be deduced by comparison of measurements of the other components.

Thus the volume of river flow available for water resource development may be estimated from the other components of the water balance, or may be measured directly and the precision of the estimate confirmed by comparison. The use of the water balance is perhaps the most powerful tool available to the hydrologist. It may be expressed by the equation:

$$R + I = E + Q + \delta S + \delta G$$

where R is rainfall, I inflow, E evaporation, Q outflow, and δS and δG changes in soil moisture and groundwater storage, respectively. It may be applied to a whole basin, or part of a basin, and may be adapted to describe the water balance of a subcomponent of the system, for instance a groundwater system.

TYPE OF WATER RESOURCES PROJECT

The need for hydrological records to assess the potential of a project or of an area depends partly on the type and scale of the likely exploitation of the water resource. A water resources project may include water supply for domestic, industrial or environmental use, for irrigation, hydroelectric power production, or flood control; each will require long-term river flow records to estimate water availability and for the hydraulic design of works.

Some types of project can be predicted further in advance and therefore have flow stations installed in good time to provide adequate data for analysis. For example, domestic water supply demands can be forecast reasonably accurately from population statistics and historic or theoretical estimates of water use; the sources which are likely to provide the required supply can therefore be monitored for a reasonable period in advance of the need for detailed design.

Water supply and river flows

In terms of the three main variables in surface water studies, which are rainfall, evaporation and river flow, river flows are the most directly applicable to project evaluation. Flow records are important for almost all forms of water supply; however, different types of water resources project have varying scales of consumption, and therefore different priorities in terms of flow series. Domestic water requirements are likely to be relatively small, and may in some cases be met by direct abstraction from a river source without the need for reservoir storage. In this case the hydrological requirement in terms of quantity may be simply an estimate of the magnitude of low flows of a given frequency at the abstraction site. The demand is likely to have been foreseen, so that analysis may be limited to an examination of a series of low flow records. The abstraction, which can be withdrawn from the river with a given reliability, can then be compared directly with the water demand characteristics required. The location and scale of water supply demand may be more difficult to predict well in advance in developing countries. This is especially true when industrial demand is a major component. The development of a new industry is likely to result in a larger change in water demand than the growth of domestic supply and may require reservoir storage near the demand site. The locations of potential industrial demand are less easy to predict, especially in the case of mining developments which depend on the results of geological exploration; thus the availability of nearby river flow records cannot be assumed. In the case of northeast Botswana in 1968, the development of important mining potential in the newly independent country depended on the potential infrastructure and particularly on the availability of a significant volume of water supply. The only information on which the urgent choice of location of a major reservoir scheme could be based was the short record of three gauging weirs on ephemeral rivers, supported by rainfall records from which an isohyetal map could be derived. Although the site chosen for the reservoir turned out later to provide a key element in the water resources development of the country, and thus demonstrated that the sites for the gauging weirs had been well chosen, the lack of a more comprehensive network could have proved an obstacle to the economic development of the country.

Irrigation demand

The demands of water for crop irrigation, especially in arid or semiarid countries where irrigation is essential for agriculture, are likely to be of a magnitude to require reservoir storage, especially as demands are liable to increase during the dry season when river flows are low. The estimation of reservoir yields through operational modelling requires relatively long-term flow records and the estimation of annual flow volumes and their variability becomes important. This requires the maintenance of a river gauge network measuring the full range of flows over a period sufficient to sample possible variations. The volumes of water consumed in irrigation are relatively large and therefore the relations between reservoir volume and yield are crucial, though it may be reasonable to accept a higher risk of shortage than for domestic or industrial supply as planting programmes tend to be more flexible and may be adjusted in periods of drought.

Hydroelectric power

An important use of river flows in many areas is the production of hydroelectric power. The need to maintain power supplies and operating head during periods of drought usually requires the use of reservoir storage. Even in the exceptionally favourable case of the Owen Falls dam at the Nile outlet of Lake Victoria, the firm power could be increased by an element of upstream storage. One advantage of hydroelectric power is that the consumptive use of water is limited to the evaporation from the storage reservoir, and this may often be offset by rainfall on the reservoir. The study of hydroelectric potential requires an extensive network of river gauges which is adequate to estimate flows at sites which would be difficult to predict without an initial survey of potential dam sites on suitable rivers.

At the time of a countrywide investigation of the hydroelectric potential of Guinea, a network of river flow gauges had been maintained over periods of 30 years or more. A regional analysis of river flows, supported by water balance studies, enabled reasonable estimates of the potential yield of a number of sites to be made. When a similar regional study of Botswana was carried out, it was possible to estimate the storage potential of a number of dam sites selected from air photography, but only preliminary estimates of the flow potential could be made without flow records. It was necessary to recommend that flow measurements should be initiated at likely sites in order that estimates could be based on actual records in the future.

Sudan example

The water demands for these broad categories of use may be illustrated by estimates of the volumes of water required for town supply, hydroelectric power and crop irrigation in the Sudan, where the semiarid climate increases the dominance of irrigation use (Sutcliffe & Lazenby, 1990). The water supply for the major city complex of Khartoum, whose population is about two million, may be estimated on a conservative scale of 100 l day^{-1} per person, as about 70 $m^3 \times 10^6$ $year^{-1}$. Although most of the country's public electricity demand is met by hydroelectric power from three dams—Roseires, Sennar and Khashm el Girba—with a total reservoir area of 700 km^2, the net evaporation loss of some 500 $m^3 \times 10^6$ $year^{-1}$ can be largely attributed to irrigation, which dominates the role of the dams. The water requirement for the principal irrigation schemes of 1.6×10^6 ha is estimated as 14 000 $m^3 \times 10^6$ $year^{-1}$, though the theoretical potential demand, if the water were available for further development, is much higher.

It is evident from this example that irrigation is by far the largest user of water in the Sudan, at 200 times the urban water supply. Indeed the potential irrigation demand exceeds the available water resources in many semiarid countries. Hence the need in many countries is for sufficient river flow records to estimate the available resources of the whole country or a region, without losing the long-term records at key sites to assess the potential of individual projects.

NEED FOR CONSISTENT RECORDS

It is therefore unfortunate that recent studies have confirmed that in a large number of countries, in Africa for instance, the number of maintained river flow gauging stations has been falling in recent years, as the funds required for the fieldwork necessary to maintain the accuracy of records have not been available. This is particularly important in view of the likelihood that climate change will be most evident in its effect on river flows; the effect of changes in rainfall or in temperature and thus evaporation will be magnified in their effect on river flows because of the sensitivity of the runoff process. It is not easy to estimate the precise value of the length of flow records in the optimal design of water resources projects, but an example from Botswana is given in Chapter 8 of the way in which the precision of flow estimates improves as the record length increases.

For the hydraulic design of most water resource projects, and indeed of all engineering projects within the river flood plain, estimates of peak river flows of different frequencies are needed. These may be derived by direct statistical analysis of flow records at the site or over the surrounding region. Other techniques of estimation of flood potential include combined analysis of rainfall and basin rainfall/runoff potential. All these require adequate networks of flow records, calibrated to give reasonable estimates of high flows. Long historical records (e.g. Herschy, 2003) are particularly valuable in studies of flood frequencies, and significant floods have been recorded in a number of countries where high floods have been observed over long periods. In China, for example, reliance is placed on historical records for flood design. Details of floods on the Chiang Jiang (Yangtze) have been preserved over 800 years by flood marks, at sites where it is reasonable to assume that the channel control has been reasonably stable.

RAINFALL

In the assessment of regional water resources, rainfall records, including snow in some areas, are an essential aid in determining the variation of the water balance over the project area. Raingauge measurements can provide comparable point information from a widespread network. This evidence is perhaps easier to interpret than the data from river flow stations with different basin areas, geological conditions and periods of record. The rainfall records can be translated into an isohyetal map which will demonstrate trends in average rainfall across a region, or the effects of topography in increasing the rainfall over high ground. A number of examples in Chapter 9 illustrate the value of rainfall data to extend the information provided by a river flow network.

Because rainfall records are simpler to collect, and cost less than a river flow network, they often cover longer periods than flow records. They are also less complex to operate and thus more likely to be reliable, especially when a group of stations can be compared. Thus flow records can be extended over longer and more critical periods in combination with rainfall records.

Rainfall data are also needed to estimate net irrigation requirements by comparison with potential crop transpiration estimates. In reservoir trials, estimates of rainfall are required to derive net evaporation from measurements or estimates of open water evaporation. Rainfall records may be useful in assessing the flood potential of the basin above the project site. Rainfall data may be used to estimate the basin storm of a given frequency, or even the potential maximum storm. Comparisons of rainfall depths from specific storms with the resulting flow hydrograph enable rainfall statistics to be converted to flow estimates.

EVAPORATION

Evaporation and transpiration measurements or estimates are a vital component of water balance studies, as they can be used to estimate the depth of rainfall which is returned to the atmosphere rather than available for runoff. A water balance analysis including evaporation is normally a vital check on an estimate of measured river flows.

Reservoir evaporation estimates are also important in the design of water resource projects. Crop irrigation requirements are related to climatic evaporation as well as soil moisture storage, and therefore evaporation rates can be used to estimate water requirements for irrigation projects. However, evaporation can often be the most difficult of the water balance components to estimate.

FURTHER DATA REQUIREMENTS

The detailed design of water resources projects requires other, more specialized, data like the suspended sediment load and bed load of rivers, in order to estimate the rate of sediment deposition and hence deduce the useful life of reservoirs. Water quality data are necessary to judge the suitability of water sources for domestic, industrial or irrigation use and environmental data are required to protect downstream habitats from serious degradation.

Groundwater investigations must be based on a wide variety of measurements or estimates, in addition to a basic knowledge of the geological structure of the study area. Requirements include a survey of depths to static water level from well information, which can be used to compile a water table or piezometric map. Fluctuations of water level over a long period can give information about recharge processes and rates. The physical properties of aquifer material (permeability and storage coefficient) can be assessed from laboratory tests or from *in situ* pumping tests, and these can be used to estimate the rate of flow through the aquifer and the water equivalent of seasonal recharge. However, these properties are best described in the context of specific studies, so discussion is deferred to Chapter 6.

CONCLUSION

Water resources projects are very variable in character and scale. They require different types of hydrological investigation and supporting information; examples will be described throughout this book. However, all these studies have in common that they are based on a need to understand the physical processes at work upstream of the project site. Although statistical techniques may be appropriate to estimate the magnitude of, for example, floods or low flows at a site, the technique should never take over from the attempt to follow the physical evidence.

2 NETWORK DESIGN AND APPRAISAL

Because hydrological networks, whether concerned with rainfall, evaporation or river flow, have much in common in the way they should be designed and analysed, it is convenient to discuss them in this chapter rather than in separate chapters dealing with specific aspects of the hydrological cycle.

It should be emphasized that advice on hydrological data collection and siting of instruments is not included in this book. These topics are discussed in many basic texts, and sound practical advice is provided by WMO (1994) and Gunston (1998).

SAMPLING OF VARIATIONS IN TIME AND SPACE

Guidelines exist for the numbers of hydrological gauges required in an area or basin, expressed as densities, or numbers of raingauges and river flow gauges per 100 km^2. However, Moss (1982) argues that general criteria cannot be specified for the design of a network, and that these should vary between different processes, geographical contexts, and levels of water resources development. Few studies have compared the values of networks with their cost, and in practice network development grows from a minimum network in the light of correlations obtained in use.

The estimation of the water resources available at a given site on a river, or over an area containing a number of rivers, depends on the sampling and subsequent analysis of flow variations over both time and space. Thus hydrological networks should be designed to be adequate to monitor variations in time and also differences over an area. In developing countries the planning of national networks must be based on the assumption that long-term records will not be available at every project site. The networks should be planned in accordance with the location of likely water resources demands, but should also cater for the unexpected development. The location of urban demand can be predicted from population densities and trends, but the future locations of industrial demand may require fuller economic analysis. The assessment of hydroelectric potential will require a more complete monitoring of the river network, and may need to be based on a preliminary assessment of potential sites. In countries where irrigation forms an important sector of the economy, the scale of demand may require monitoring of all major rivers; however, in more arid countries where irrigation is vital, the number of significant rivers may be limited. This is the case in the Sudan, where the key sites depend strongly on the river network. The criteria for gauge density can benefit from statistical analysis; the cross-correlations obtained within the initial network will indicate the number of gauges required, consistent with the economic development of the country.

Type of variation

Between rainfall, evaporation and river flow, variations in time may most easily be detected in rainfall and river flows. Although rainfall stations are easier to operate, a single record may not fully represent the variability of available water over a basin because of measurement errors, random incidence and altitude effects. It is usually necessary to combine a number of records to monitor variations over time, whether over seasonal, annual or longer time scales.

On the other hand, river flow measurements are more complex and expensive to initiate and maintain, as a gauging station must be constructed and monitored, with usually the need to calibrate the station by regular measurements. However, because river flows are the residual of the water balance process, they are more sensitive to variations in water availability and therefore a series of flow records will make it relatively easy to detect changes over the years. Indeed, the outflows from Lake Victoria in East Africa are so sensitive to relatively small changes in rainfall that some hydrologists have not accepted the validity of the flow records and have put forward a variety of alternative explanations.

Moreover, river flows are the main subject of most water resources studies, so that variations are more directly relevant to the long-term yield of projects and direct measurement provides the most accurate estimate. The advantage of rainfall networks is that their length of record is in general longer than those of river flow stations. However, once estimates of areal rainfall have been derived by combining a number of records, the time series need to be related to river flows by some form of rainfall–runoff model.

Variations in time

The geographical distribution of gauging stations is not the only consideration. The sampling of variations of river flows with time over a region has become more important with increased knowledge of the extent of periodic droughts or fluctuation of supply over wide parts of Africa, for example. It is increasingly understood that the climate and water resources of large regions can vary between successive decades or longer periods. Whether these variations are random, cyclical or result from climate change, it is important to monitor them and thus to assess water resources potential over as long a period as possible. It is therefore extremely important for rainfall and river flow stations to maintain complete and consistent measurements at a number of core stations which can be used to sample variations with time. These basic stations can be used to extend shorter records by correlation to estimate time variations at the short-term stations; therefore the correlation matrix between stations is an important indication of the consistency of rainfall and river flow and thus the density of stations required.

Although measurements or estimates of potential evaporation can be made at a number of sites, it is generally accepted that evaporation varies much less from year to year and from site to site than other hydrological variables. Although seasonal fluctuations are significant, evaporation appears to be much more conservative than other variables and rather fewer observations are needed to sample the variations.

Variations in space

Estimates of spatial variations of water resources are particularly important where the resources of an area are being assessed, or where there is a choice between different project locations. The primary criterion for the viability of a project is usually the mean and variability of river flows at or near the site. Therefore, the hydrological network should be designed to sample the variation of runoff over a region which may exhibit differences in climate and geology, and thus enable the hydrologist to estimate the water available at a number of ungauged sites within the region.

Because a dense network of long-term river flow stations would be expensive to establish and to maintain, it is preferable to use a network of rainfall stations and flow stations to monitor the variation of water resources over a region. Indeed, it could be argued that it is easier to compare rainfall records over an area in order to deduce regional trends, because the records are not confused by variations of basin area, geology, and thus runoff coefficient, which complicate comparisons of flow records. It is essential to support the spatial distribution of runoff with that of rainfall. Although rainfall distribution is subject to influences like distance from the moisture source and the orographic effects of topography, the rainfall depths are more easily comparable than runoff depths. The spatial distribution of rainfall can be more easily mapped than that of runoff, and it is common to describe this by means of an isohyetal map, or preferably to map net rainfall, and then to relate runoff depth to this map. It follows that a combination of long-term and short-term stations is necessary to define the spatial distribution of rainfall.

In fact a combination of the records of rainfall, evaporation and runoff is also necessary to understand the water balance of an area, and therefore all three hydrological components need to be monitored by a network of measuring stations. The links between these components may be treated as simple comparisons, or compared by statistical comparisons or by conceptual modelling, but at this stage we are considering the planning, operation and monitoring of the different networks.

QUALITY APPRAISAL

If hydrological records are to be used in assessing available water resources over a region, it is of course necessary that they should be as accurate as possible, but it is probably more important that they should be consistent, both over time and from one site to another. Recent short-term flow records at a project site are often extended to cover a longer and more critical period, using long-term rainfall or flow records in the vicinity. Bias will be introduced if the recent records at the long-term sites have been measured or processed differently from the earlier records, even if the recent records are in fact more accurate as a result of the change. It is therefore essential that reviews of hydrological records should be applied consistently over the whole period.

The appraisal of rainfall and evaporation records must start with the investigation of sites and measuring techniques. These may change over the years even if the record is published as a continuous series. However, changes are usually recorded in station files or annual reports and may require adjustments to the basic series. Double-mass or cumulative comparisons over time between single stations and either adjacent stations or groups of stations can reveal the dates and directions of significant changes which require detailed investigation.

Accuracy and precision

An example of the way in which consistent records may be more important than precision is illustrated by the problem of the water balance of Lake Victoria. The coverage of raingauges around the lake and its basin has improved considerably over the years, especially since the establishment of gauges on islands within the lake after 1967. However, the contribution of the heavy rainfall of 1961–1964, which led to a significant rise in the lake, was more easily reproduced (Piper *et al.*, 1986) by using the small number of long-term gauges around the lake, after adjusting for site changes, than by using the whole increasing but non-homogeneous network. This problem is discussed in more detail in Chapter 8.

RATING CURVES

River flow stations are more complex to maintain as a homogeneous record, even though standard measurement techniques are used. At sites where river level records have been collected over a long period, the quality of the derived flow records is likely to depend on the use of discharge measurements to derive rating curves, or curves relating water level and discharge, at the site. The factors affecting the quality of flow records include the range of gaugings compared with the range of level and flows experienced at the site, the methods used over the years to derive successive rating curves, and the stability of the site itself in terms of control. For example, at gauging stations with sand beds in Botswana, the river bed may change shape dramatically after each significant flow event.

The overall quality and consistency of the record can be assessed by using all, or samples of, the historical discharge measurements to derive rating curves by consistent computer-based methods for the whole period of records, divided if necessary into relatively stable periods. The advantage of treating all the gaugings at a site in one comparison is that a more consistent set of rating curves is likely to result. In any single year of gaugings, it is not likely that sufficient high floods will have been measured to define the rating curve precisely for high flows; the upper part of the calibration is normally defined by channel control which varies little from year to year, and thus gaugings from a longer period can be combined to derive a high flow rating in a form compatible with the geometry of the site. For example, in a study of about 40 years' flow on the Nile below Lake Kyoga in Uganda, it was evident that in individual years the range of levels and flows was inadequate to derive a precise rating, as the flows were highly damped by storage in Lake Victoria and Lake Kyoga. However, analysis of all gaugings over a period allowed a family of curves to be derived, which allowed for changes in control level from year to year. In addition to minor annual changes in control, there was a major shift in the rating curve after the rise in Lake Victoria level and outflow after 1961; both these changes could be incorporated in the rating curves (see Fig. 7.4).

In the initial stages of the investigations leading to the *Floods Studies Report* (NERC, 1975) for the British Isles, a countrywide investigation was made of all the gauges with at least five years of record; the rating curves were compared with high flow gaugings and consistent curves were derived, either for the whole period or for shorter periods. In addition to ratings which were used to derive flood flows from peak levels, an assessment of the quality of each flood record was made on the basis of the precision and stability of these ratings and other factors.

The use of measurements from single years may produce high flow ratings based on extrapolation which are inconsistent from year to year, and in some countries may determine a high proportion of the annual runoff. On the other hand, low flow ratings

Fig. 2.1 *Rating changes for Bahr el Jebel at Mongalla; flow corresponding to the gauge level of 12.25 m.*

at sites with mobile beds are likely to change from flood to flood; the simultaneous study of a number of years of measurements should make it possible to date changes and to monitor progressive movements of bed controls. An example of such progressive changes is illustrated in Fig. 2.1, where successive discharges of the Bahr el Jebel at Mongalla in the southern Sudan, corresponding to a gauge level of 12.25 m, reveal changes of rating which are consistent with measured bed levels.

Assessing rating curves

The direct assessment of ratings can be supported by indirect techniques like double-mass analysis and inter-station correlation. The double-mass comparison of flow records at upstream and downstream sites on the same river, or draining adjacent basins, should reveal a linear relation with minor variations, and the dates of any changes of slope should draw attention to the dates for detailed investigation of critical non-homogeneity.

An example is given in Fig. 2.2. The cumulative annual flows of the Bahr el Jebel at Mongalla are compared with outflows from Lake Victoria at Jinja, where the natural and stable rating curve at the Ripon Falls, linking lake level and outflow, has been maintained since the construction of the Owen Falls dam at the outfall. There are other tributaries between the two sites, which are about 900 km apart, but the relationship between the two sites has been reasonably stable; the flows at Mongalla may have been over-estimated in the early years before regular discharge measurements began. Inter-station correlations of annual flows should reveal stations to which particular attention should be paid. Even simple comparisons of rainfall and runoff depth over a region can identify stations whose records merit particular attention; such a comparison in the central mountains of Sri Lanka (Fig. 8.2) showed that most stations were reasonable but that at least one was an outlier in terms of water balance.

If these or other analyses show that there are likely to have been faults in the original rating curves, these can be improved retrospectively by assessing the whole

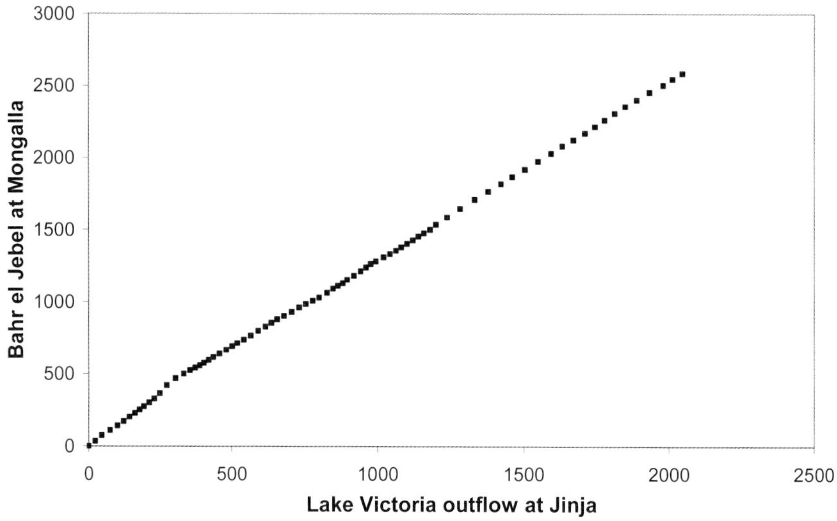

Fig. 2.2 *Double mass curve: cumulative annual flows (km^3) of the Bahr el Jebel at Mongalla vs outflows from Lake Victoria at Jinja, 1905–1982.*

period of discharge measurements, provided that the sites are either reasonably stable or progressive in their changes.

COMPARISON OF RAINFALL AND RUNOFF

Comparisons of river flow series with the corresponding series of basin rainfall or preferably net rainfall can also be used to monitor or improve flow record quality. For example, a single abrupt change in the relation between stations or between rainfall and runoff might be explained by a simple unrecorded change of gauge zero. During an investigation of hydroelectric potential in the volcanic area of the North Island of New Zealand (Sutcliffe & Rangeley, 1960a), the only river gauge was that of the Tongariro at Turangi, which had been used for subjective forecasting for hydroelectric power generation. It was known that the gauge zero had been changed by a misplacement of the float connection, but the precise date was not known. Cumulative differences between basin rainfall and runoff depth were used to derive a time series of apparent basin loss (Fig. 2.3), and the date of the change, June 1952, was revealed by this simple comparison and the flow data adjusted before analysis.

USE OF RECORD NETWORK IN TIME AND SPACE

Once the networks of rainfall and runoff records have been appraised, and adjusted where necessary, they may be used to extend short-term flow records by correlation with long-term flow or rainfall records. Alternatively, a comparison of flow records for a standard period with rainfall distribution over the region can lead to an assessment of the water resources of an area. As will be seen in later chapters, these extensions and regional studies are much easier to carry out in humid rather than arid climates.

At this stage it is appropriate to discuss the influence of their use on the number of gauges required. The numbers of rainfall and flow gauges required for water resources assessment cannot be linked simply to area, though tables of this type may be a useful

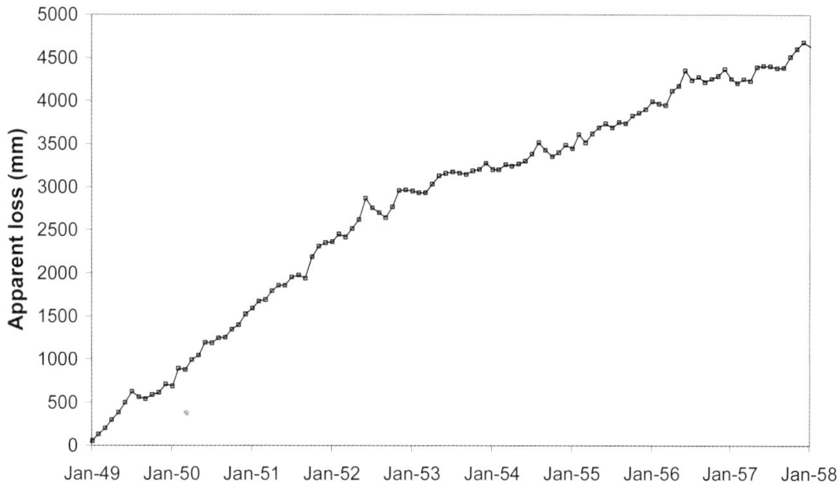

***Fig. 2.3** River Tongariro at Turangi: cumulative apparent basin loss (1949–1957) (after Sutcliffe & Lazenby, 1990).*

preliminary indication. Factors to be taken into account include topography, population density and the degree of economic development, and the variability over the region of climate and geology. For example, in arid regions with few major rivers, e.g. Sudan or Botswana, flows may be measured at key sites without a high density of gauges. The centres of population and likely water resources requirements must be taken into account in network planning, but higher gauge densities will be required if the areal variations of rainfall and geology make runoff from ungauged basins difficult to predict. Correlation analyses on basin characteristics can be used to assess this predictability and indicate the density required, in the same way as correlation between gauges can be used to assess the period of extended record needed to estimate mean runoff with a given precision. However, there must in most cases be a compromise between the optimal network judged on hydrological criteria and the resources which are allocated by the authorities of a developing country. The gap between these may be reduced by the understanding of the importance of records in developing water resources, but the role of the hydrologist must be to provide more effective ways of collecting, appraising and processing data, and using them in resources assessment.

CONCLUSION

Whatever the climate and water balance regime of the region, and the scale of the water demand in relation to water availability, the need for hydrological records remains. The requirement for precise water resources assessment is a network of reasonably accurate long-term rainfall and river flow records which can be used to sample variability of river flows in both time and space. For this purpose consistent records are of greater importance than absolute precision, and a careful appraisal of the consistency of both rainfall and river flow records is an essential preliminary to reliable water resources assessment.

3 RAINFALL

Rainfall is the major contribution to the water balance and thus the water resources of an area. Strictly precipitation is the more inclusive and correct term, as it includes snowfall and other forms of contribution; however, nearly all the examples discussed here are taken from countries where rainfall is most common. Rainfall is the simplest term of the hydrological cycle to measure, at least in terms of daily values, and records therefore in general cover a longer period than other hydrological components. Indeed, one of the first actions of settlers in many countries seems to have been the installation of raingauges. In Sri Lanka, for example, the central mountain range was developed with tea plantations, and most sites possessed a raingauge, so that many records are continuous from the 19th century. For the same reason, and because of the lower costs of installation and operation, the raingauge network is usually denser than observations of evaporation or river flow, so that variations in water input as a function of time or over an area can most easily be studied in terms of rainfall.

In this chapter the topic of rainfall is treated in terms of its measurement and the analysis of records to derive mean values and variability according to season and from year to year; the analysis includes the assessment of maximum values. Whereas it was at one time considered sufficient to estimate the averages over standard periods, usually covering 30 years, it has recently become clear that there are significant differences in values in different periods, and changes and trends are now studied more carefully.

The spatial distribution of rainfall can be analysed by the construction of isohyetal maps, illustrating the variation of average rainfall depth with topography, which includes both position and elevation. The main hydrological objective of the isohyetal map, which can be derived subjectively by someone familiar with the effect of topography, or objectively by computer-based techniques, is to deduce the rainfall over a basin to compare with runoff depth. It can be particularly helpful where rainfall records are distributed unevenly through the basin and the influence of elevation could be underestimated. Mountainous areas tend to be under-represented by raingauges because of the difficulty of measurement and access, as in the mountains of Iran discussed later, but these areas are often particularly important for water resources inputs.

Time variations of basin rainfall can be described by the use of station or basin indices, which are related to variations from the long-term average. The number of stations required to estimate basin rainfall to a given precision will depend not only on the climate but also on the duration of the time interval, the basin area and topography, which all affect the variability of rainfall. Examples are given to illustrate the problems involved.

RAINFALL MEASUREMENT

There are many different types of rainfall gauge, but the main distinction concerns the time period over which rainfall is recorded. The standard gauge, of which for example there are some 4000 in Britain, is the daily gauge in which the rainfall caught in a funnel of a given area is extracted at the same time each day and measured in a container which gives the equivalent rainfall depth. The standard elevation above the ground varies according to country, largely according to the incidence of snow storage; the diameter of the gauge funnel also varies.

In remote areas where observers are not available, it may be necessary to supplement daily gauges with storage gauges or automatic gauges. In an investigation undertaken in 1958–1960 of the runoff potential of a mountainous area in the North Island of New Zealand, the four existing raingauges were concentrated in lower areas where reasonably long records had been collected. To investigate the effect of elevation and aspect on rainfall distribution, an additional 12 temporary gauges were installed. Half of these were located where observers could be persuaded to accept them, and the remainder were storage gauges installed and visited weekly (Plate 3.1) at key sites where there were no inhabitants. One gauge was located at an elevation of 1750 m at the top of a ski-lift on Mount Ruapehu, and supplemented by a gauge to measure snow in winter. Use was also made of one existing gauge which had been installed at a remote site; it required a two-day expedition to read this storage gauge. Automatic weather stations or storage gauges which provide regular frequent observations and daily totals would now be used in these circumstances.

Recording gauges have been used for many years to measure rainfall intensity over shorter periods than daily values. Early examples (Strangeways, 2003) used mechanical methods like tipping-bucket devices which recorded the frequency of unit rainfall volumes on charts, but more recently solid-state loggers have been developed for the same purpose. These recording gauges can be combined with daily or storage

Plate 3.1 *Storage raingauge in the Tongariro basin, New Zealand.*

gauges, or even radar observations, to derive the time distribution of storm rainfall over a basin for comparison with flow records in order to investigate the flood response of a basin to heavy storms.

Mean and variations

The simple analysis of a rainfall record provides the monthly and annual average over the period of records, and can provide a measure of variability in the form of maximum and minimum, range or standard deviation. It has been standard practice to cite averages for standard periods of 30 years, e.g. 1961–1990, which was assumed to provide a sufficient period to give a stable average, not only for the annual total but also for the monthly means. However, it has been found in some areas that variations can persist over long periods; for example, it seems that the rainfall regime over and around Lake Victoria has changed after 1961, with somewhat higher rainfall in the second, November–December, rainfall season. This has had significant effects on the outflow from the lake basin, because the mean rainfall over the lake is nearly equal to lake evaporation. Another example of long-term change in rainfall has occurred over the Sahel region of West Africa in the years since 1974, to the extent that water resources projects based on early records have fallen significantly short of their design yields. Weak links have been found between annual flows in some but not all major basins in sub-Saharan Africa, and El Niño Southern Oscillation (ENSO) effects (Farquharson & Sutcliffe, 1998). It is important to quote the period from which averages have been derived, and to treat with caution those averages which are based on limited periods of records. This is likely to become more significant in the light of predicted climate change.

Nevertheless, annual average rainfall estimates, when these have been adjusted to a common period, provide the most useful information on the distribution of available water over an area. When plotted on a topographic map which indicates elevation, these estimates provide evidence of the way in which the topography, in particular relief and aspect, influences rainfall depth. The monthly averages, especially when plotted on a map or diagram in their relative station positions, provide an instant picture of the seasonal rainfall pattern of the whole area, and will reveal any trends in seasonal distribution over the region. An example is shown in Fig. 3.1, where the rainfall over the Senegal basin (which includes parts of Mali and Guinea) is illustrated by isohyets and in Fig. 3.2 by monthly histograms. It is evident that the annual average rainfall decreases rapidly from north to south over the basin, but also that the rainfall is highly concentrated into a limited season, and further that this season decreases in duration from south to north. A time series of basin rainfall or at individual stations would show that rainfall has decreased over the years, but in fact this is demonstrated more dramatically in Chapter 9 by the river flow series.

Adjustment of short-term averages

When some of the stations have records for a consistent long-term period, for example over a 30-year period, and other stations cover shorter periods, it is necessary to deduce compatible estimates by deriving averages over a common, long-term period. This is most easily done by comparing each short-term station with one or more nearby long-term stations, either using correlation analysis or simply by adjusting the average at the

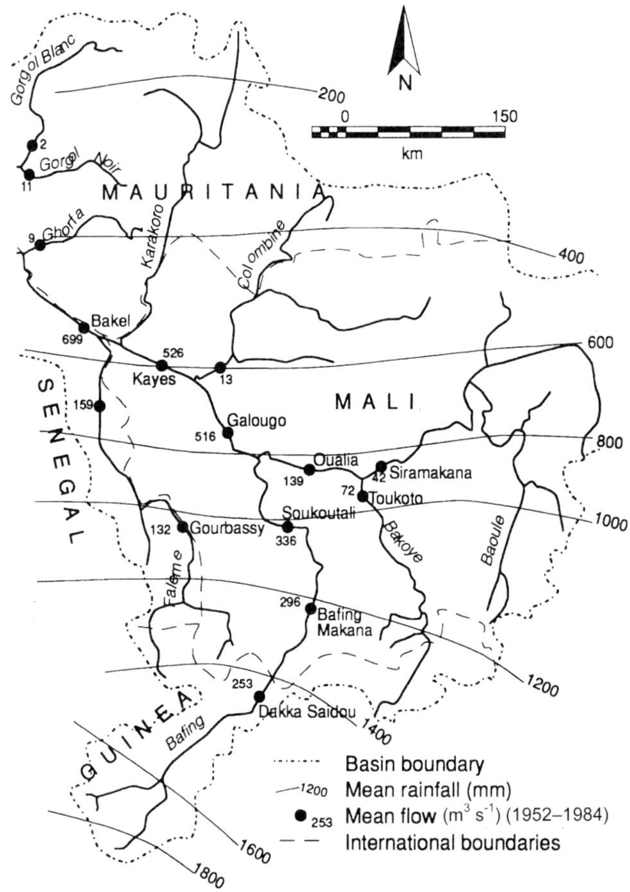

***Fig. 3.1** Annual average rainfall (1951–1980) over the Senegal basin.*

long-term station by the ratio of the short-term to the long-term station during the common period.

CASE STUDIES OF SPATIAL VARIATION

Once the rainfall data have been reduced to consistent long-term averages, the isohyetal map is the simplest means of assessing the variation of average rainfall over an area. In areas of moderate relief, the distribution will be controlled largely by position, and the drawing of the isohyets can be based largely on interpolation between station values.

Botswana

An example of this was an early study of water resources in Botswana, at a time when flow records were so sparse that rainfall data gave almost all the information on the likely volumes of runoff available at different sites. It was possible to make use of rainfall data from adjacent countries, namely South Africa and what was then Southern Rhodesia, now Zimbabwe. The combination of a limited number of records in eastern

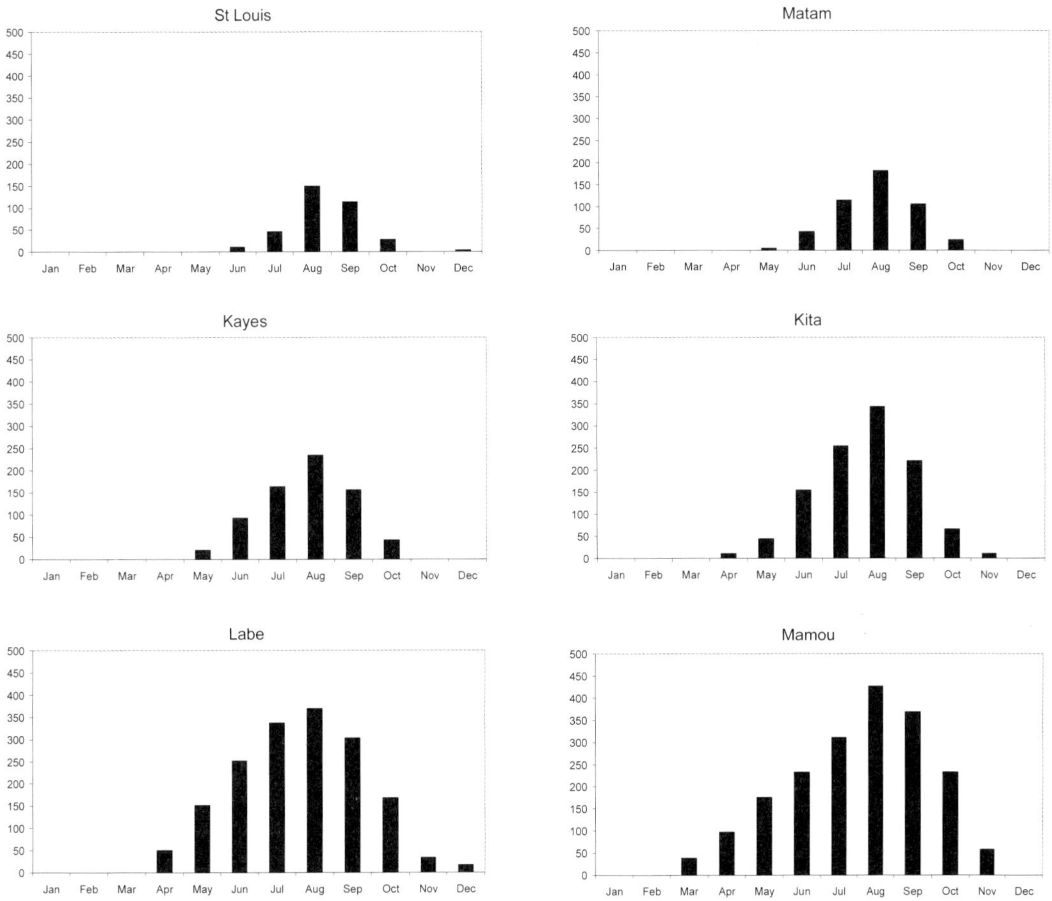

Fig. 3.2 Monthly average rainfall (mm) over the Senegal basin.

Botswana, and the adjacent isohyets, suggested the rainfall distribution illustrated for northeast Botswana in Fig. 8.7. The study indicated relatively high rainfall in the south of the country and in the hills near Francistown in the northeast. There was an indication of lower rainfall in the centre, over an area bordering the lower ground of the Limpopo, and this was supported to some extent by the vegetation distribution. In the context of the study, which was to plan the siting of a major reservoir to support the mining potential of the newly independent country, this rather slender evidence was the major factor in recommending that the reservoir should be located near Francistown, where a short record of river flows was available, rather than nearer the location of the mining development itself.

New Zealand: storm direction

In the study of hydroelectric potential in the North Island of New Zealand, referred to earlier, it was necessary to study the rainfall in some detail; the estimated average runoff, based on a short period of nine years' flow records on the main river, the Tongariro, suggested that the runoff depth exceeded the average annual basin rainfall deduced from the official countrywide isohyetal map. In this case it was necessary to

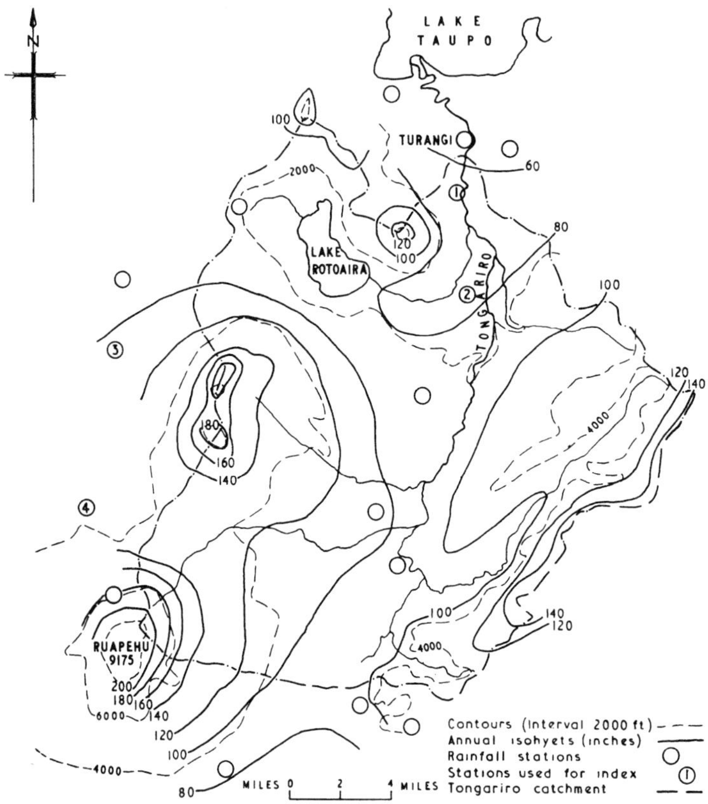

***Fig. 3.3** Tongariro above Turangi: isohyetal map (1929–1958) (after Sutcliffe & Rangeley, 1960a). Note: 1000 ft = 305 m; 100 inches = 2540 mm; 1 mile = 1.61 km.*

relate rainfall to elevation, as the topography around Mount Ruapehu (2800 m) varied considerably in elevation (Fig. 3.3) (Sutcliffe & Rangeley, 1960a). After a number of short-term stations had been established and data recorded for only about a year, it was clear that the precipitation was highly dependent on topography. Correlation of precipitation with elevation showed a steady linear increase from 1500 mm at the general elevation of 800 m up to an estimated 5000 mm at a combined raingauge and snowgauge established at the top of a ski-lift on Mount Ruapehu at 1750 m. In this particular case snow storage was considered as the area included the major ski resort of the North Island; however, a simple estimate was made of the proportion of precipitation stored as snow. This was based on monthly temperatures recorded at higher stations, adjusted at an appropriate lapse rate, to give the elevation at which the monthly mean temperature was less than 0°C. On the assumption that precipitation above this elevation would be stored as snow, this showed that only a small proportion of the basin precipitation was affected.

The records showed that the average to the north of the mountain complex was greater than to the south, although the prevailing winds were westerly. This anomaly was explained after a discussion with a local teacher, who was a keen fisherman and had observed that the heavy falls occurred during northerly storms bringing rain south over Lake Taupo. This observation was checked and confirmed by the New Zealand Meteorological Department, who kindly examined the synoptic charts during the dates of historic heavy storms. This information enabled a local isohyetal map to be drawn, with the rainfall shadow repositioned to take account of the revised storm direction. The anomaly of the apparent excess of runoff disappeared once the basin averages were revised on the basis of this map. This information was also used to correct the national isohyetal map.

Western Iran

In a study of the water resources of the rivers draining the Zagros mountains in western Iran (Sutcliffe & Carpenter, 1967), it was found that the available rainfall records, analysed after discussion with the Iranian Meteorological Service, revealed a more complex situation. The prevailing winds brought moisture from the plains of Iraq towards a series of ridges rising up to the Zagros and running from NNW to SSE. The rainfall increased up to the top of each ridge, decreased on the leeward side of the ridge, and increased again at the next ridge. A section through the mountains (Fig. 3.4) illustrates the effect of the topography on rainfall depth. Thus rainfall is related to both elevation and position. However, the complexity of the topography and the resulting rainfall distribution made it difficult to compile an isohyetal map in the usual way. After trial and error, the rainfall averages were reduced to a standard elevation of 1000 m, in the same way as atmospheric pressures are reduced to sea level, using a linear gradient of 300 mm/1000 m elevation for this purpose. Once a standardised value of rainfall at an elevation of 1000 m had been derived, it was simple to draw an isohyetal map (Fig. 3.5) of these values, which showed standardised rainfall decreasing over the project area in the direction of the prevailing winds. The rainfall at higher elevations could be deduced from the standardised rainfall using the same linear gradient; basin mean rainfall could also be estimated from basin elevation. This interpretation of the rainfall distribution was supported by the natural vegetation of the

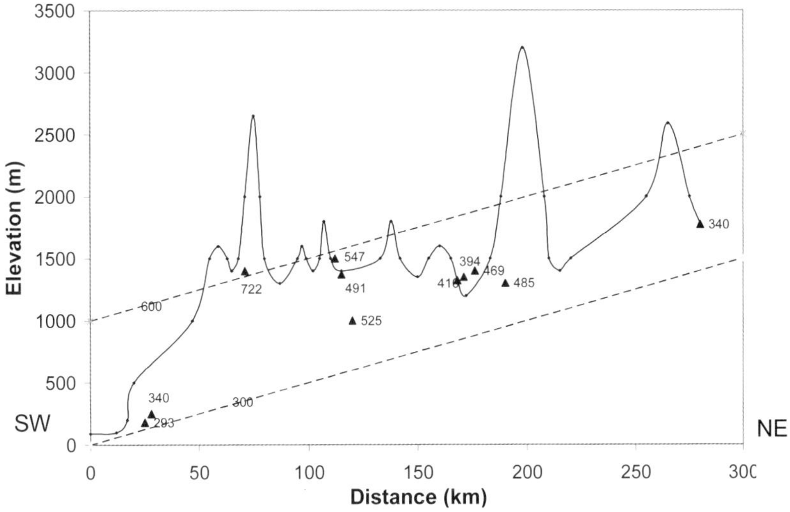

Fig. 3.4 *Zagros Mountains, Iran: cross-section southwest–northeast showing annual average rainfall (mm).*

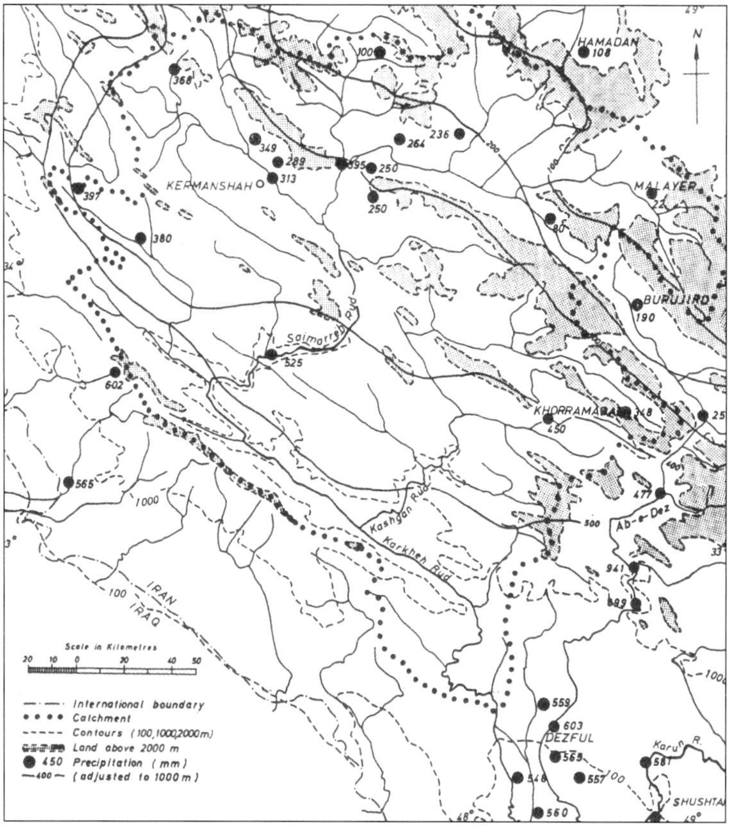

Fig. 3.5 *Zagros mountains, Iran: isohyetal map of annual average reduced rainfall, mm (1945–1964) (from Sutcliffe & Carpenter, 1967).*

Plate 3.2 *Oak scrub on the southwestern foothills of the Zagros mountains, Iran.*

mountains (Plate 3.2); oak woodland and scrub were observed to be thicker on the west-facing slopes above a given elevation, but became less dominant towards the east after the first or second ridges. It was also noted that the woodland was thicker on north-facing slopes, where the insolation and evaporation would be lower. In order to deduce the average rainfall over a given basin, the linear gradient had the advantage that the average standardised rainfall could be deduced from the isohyetal map, and then simply adjusted to the mean elevation of the basin.

This illustrates a theme which will recur in this book; it is unsafe to draw conclusions without visiting the study area and discussing the regime with those who know the area well. These examples demonstrate that there is no substitute for first-hand knowledge of the project area when water balance studies are being carried out. Discussions with local observers and meteorologists are also essential to avoid drawing inaccurate conclusions. Isohyets can be derived by statistical methods, in which rainfall is related to position, elevation, aspect and other variables like distance to the sea or a barrier; however, some understanding of the factors affecting the local meteorology and even the natural vegetation should be included in the analysis.

Northern Iran: use of vegetation evidence

The value of the evidence from natural vegetation is illustrated by the water balance of the Alborz mountains between Tehran and the Mazanderan plain near the Caspian Sea. The area was investigated in 1968 (Sutcliffe & Swan, 1970) in the context of transfer of water to the south to meet growing demand. There were striking differences between the two aspects of the Alborz massif in the seasonal distribution of rainfall (see Chapter 8) and the vegetation. To the south the rainfall is low, and confined largely to the months October to May; to the north of the mountains the rainfall is higher and occurs throughout the year, especially in the foothills. The rainfall distribution is illustrated in Fig. 3.6. The difference in vegetation is equally striking, with thick forest and woodland

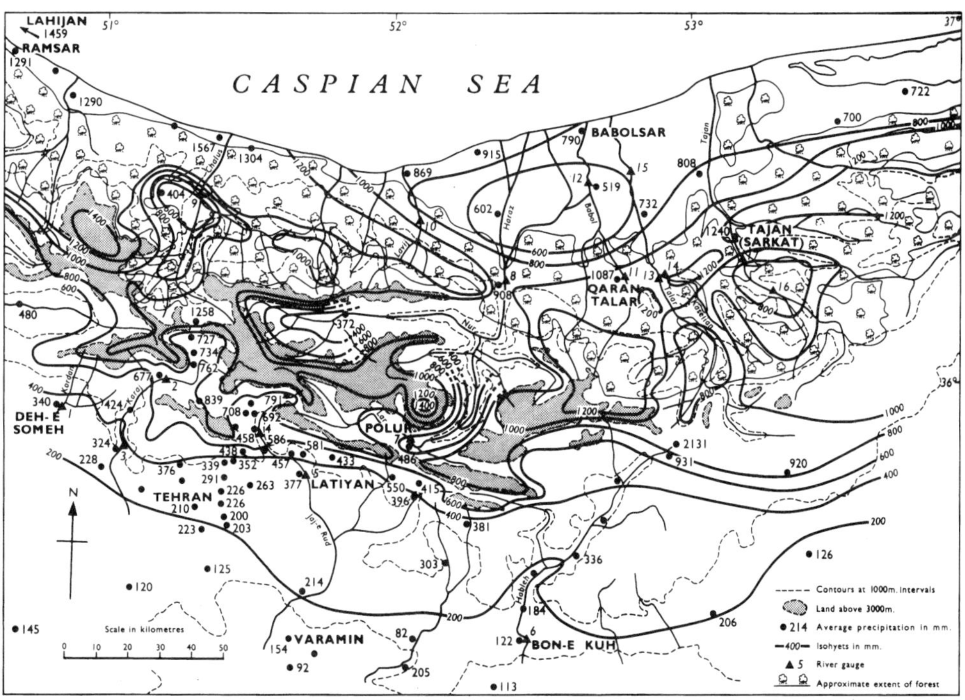

Fig. 3.6 Alborz mountains, Iran: isohyetal map of annual average rainfall, mm (1951–1966) (from Sutcliffe & Swan, 1970).

(a)

(b)

Plate 3.3 *(a) Woodland and rice cultivation on the northern slopes of the Alborz mountains, Iran; (b) the southern slopes of the same mountain range.*

(Plate 3.3(a)) to the north of the mountains and little vegetation (Plate 3.3(b)) to the south, except for a brief period in spring.

Point estimates of mean annual rainfall were compared with the natural vegetation during a tour of the area, supplemented by a map with an outline of the forested area; it was evident that the woodland coincided with the area of perennial rainfall and that the density of the vegetation was related to the depth of annual rainfall. The thickest vegetation was found on the nearest ridge facing the coast, and thinned out to scrub towards the south at a rate which depended on the prominence of the ridges. Towards the watershed, the south-facing slopes become sparse and then bare. Thick woodland extended further south in those valleys, like the Babol, which were open towards the coast.

It was evident that the woodland is limited to those areas near the Caspian Sea with summer rainfall, where the trees are able to survive the summer with its higher transpiration, and that the density depends largely on the total summer rainfall. Within the foothills, where the rainfall is seasonally more uniform, the woodland density indicates the total rainfall which in turn is controlled by the ridges to the south of the coastal plain.

Local meteorologists explained that though heavy frontal rain often passes from west to east over the Caspian side of the mountains when rain is also occurring to the south of the range, the rainfall confined to the northern slopes of the foothills occurs with northerly winds from the Caspian Sea at relatively low levels. This explains the shielding effect of the ridges and the complex relation between topography and rainfall in the northern Alborz. The relation between precipitation and elevation was more direct to the south of the mountains.

Although the number of gauges was small, comparison of rainfall records with observations of the natural vegetation led to the compilation of an isohyetal map (Fig. 3.6) on which the station averages and woodland distribution are included. Once an isohyetal map has been compiled, the long-term mean rainfall over a basin or a number of basins can be deduced. This average basin rainfall is the primary factor in the water balance of an area, together with the monthly distribution of rainfall. This map formed the basis for a water balance of the north- and south-flowing rivers (see Chapter 8), which, because of the differences in actual evaporation, including interception by the woodland canopy, carried similar depths of runoff.

VARIATION OF RAINFALL OVER TIME

Because rainfall records often cover a longer period than river flows, the rainfall series is likely to provide more information on variations in water resources than runoff series alone. It is often useful to relate river flows with basin rainfall over the common short-term period, and then to extend the flow record using this relation and the longer rainfall record. For this purpose it is necessary to derive a consistent series to represent the rainfall over the basin during both the short-term and long-term period. The evidence which can be used for this purpose is the set of monthly rainfall records at a number of sites within or near the basin. It is unlikely that the records will be complete at all stations over the whole period, and therefore the method of deriving the series must remain reasonably consistent.

Tana basin, Kenya

An example of this problem was presented during a study of the water resources of the upper Tana basin in Kenya (Sutcliffe & Rangeley, 1960a). The basin of 9300 km^2 is bounded by the Aberdare mountains and Mount Kenya, and ranges from moorland at over 3000 m to semiarid plains at about 1000 m. As the basin is near the Equator the rainfall pattern is markedly seasonal with rainfall occurring in April–May and November–December. Average rainfall varied from 650 mm on the plain to 2000 mm at about 2500 m in the mountains, with a rainfall shadow between the two ranges. All of the 166 rainfall records were used to compile an isohyetal map (Fig. 3.7) from which the long-term average basin rainfall was estimated as 1170 mm. This compared with an average runoff of 293 mm, but the runoff estimate was based on only 11 years of record at the time.

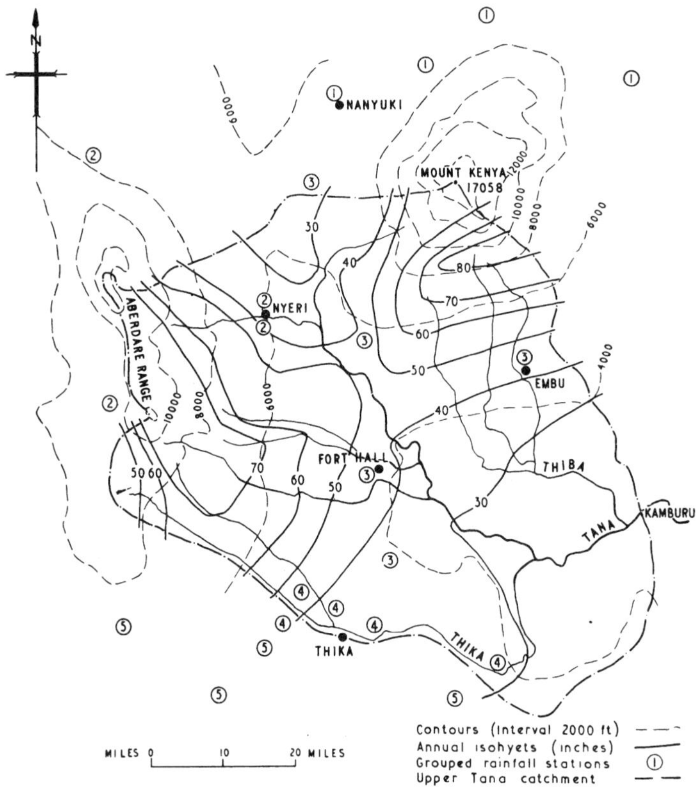

Fig. 3.7 *River Tana, Kenya: isohyetal map of annual average rainfall (1928–1957) (from Sutcliffe & Rangeley, 1960a). Note: 1000 ft = 305 m; 100 inches = 2540 mm; 1 mile = 1.61 km.*

As 22 rainfall stations were available for a common period of 30 years (1928–1957), their averages were computed over this period, and each annual rainfall total was expressed as a percentage of this station mean to give a station index for each year. The stations were divided into five groups according to location, and the mean of the station indices was averaged for each group. The mean of all the station indices was taken as the catchment index, and as a test of homogeneity the group indices were correlated with the catchment index; coefficients varied from 0.88 to 0.97. This suggested that the catchment index was representative of variations of annual basin rainfall. As a reasonable number of records dated back to 1908, reducing from 12 in 1916 to five in 1908, a catchment index was calculated back to this date. It was used to estimate a 50-year record for comparison with measured river flows. This method of deriving a rainfall index is often referred to as an isopercentile approach, as the rainfall can be expressed as a percentage of the long-term mean. It has the advantage that it can be used to compile an index from a raingauge network with missing data, as it assumes that the percentage variation from the mean can be estimated from a varying number of gauges.

Betwa basin, India

A similar approach was used in a research study in the Betwa basin near Bhopal in central India. An analysis of the surface water balance (Sutcliffe *et al.*, 1981) was

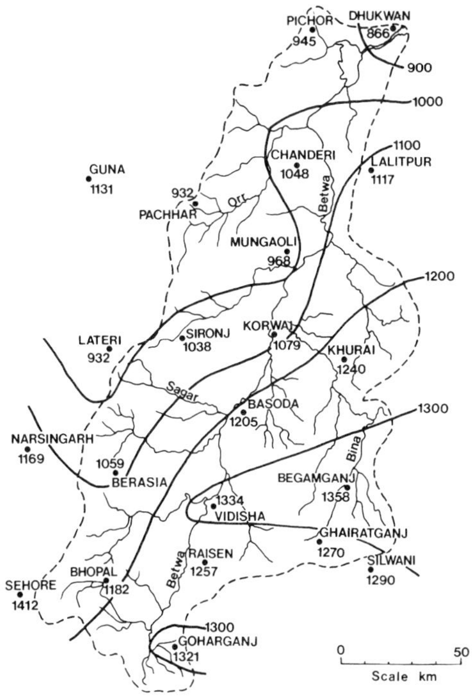

Fig. 3.8 Betwa basin, India: isohyetal map of annual average rainfall, 1926–1975 (mm) (from Sutcliffe et al., 1981).

required to support a research investigation of groundwater recharge in a typical hard-rock area of Deccan trap basalts, sandstone and granite. Over the basin of 20 600 km^2 the elevation only varies between about 300 and 700 m, but the average rainfall decreases from south to north from over 1300 mm to less than 900 mm. The average rainfall over the whole basin was estimated from an isohyetal map (Fig. 3.8) for the 50-year period 1926–1975 as 1138 mm. The rainfall distribution is highly seasonal, with 92% of the total rainfall occurring in the four months June–September. In order to derive a seasonal balance, the monthly rainfall series at each station was converted to percentages of its long-term mean annual total, and the average of 17 monthly station indices were averaged and multiplied by the basin long-term average to give a monthly series.

It is interesting to note the difference between the variability of station and basin annual rainfall totals. The coefficient of variation (CV, the standard deviation divided by the mean) of rainfall over the basin is 18.5%, compared with the CV at individual long-term stations which ranges from 25.3 to 34.1%. This illustrates that rainfall over an area is much less variable than rainfall at a point.

Short-term rainfall

A similar technique was used to derive short-term rainfall over the basins upstream of gauging stations in Britain during the investigation leading to the *Flood Studies Report* (NERC, 1975). In comparing storm distribution over relatively small basins with the resulting runoff in order to derive unit hydrographs at different sites, the total rainfall over the storm was calculated by the isopercentile method and related to the long-term basin average. The ratio of the storm total at each station to the station annual average was expressed as a percentage and averaged for all the chosen daily gauges; this percentage was then multiplied by the annual average for the basin, assessed from large-scale isohyetal maps. This was distributed through time in accordance with the hourly falls recorded by one or more autographic gauges, weighted where necessary according to the distance from the centre of the basin. This procedure gave a short-term rainfall series for each storm which could be compared with the resulting runoff, to deduce the response of the basin.

PRECISION OF ESTIMATE OF BASIN RAINFALL

When extending river flows by comparison with basin rainfall, it is important to estimate the number of gauges required to estimate the basin rainfall to a given precision. This may be based on statistical analysis of existing networks. Some estimates have been based on the assumption that point measurements are independent measurements of the mean, and thus that the standard error of the basin estimate is the standard deviation of the point estimates, divided by the square root of the number of stations. This neglects the underlying effect of topography on rainfall, and exaggerates the number of gauges required. The true precision can be deduced by separating the variability of rainfall totals into that due to position, which may be termed the "persistence factor", and that due to random distribution; it is the latter which affects the precision of flow extension.

The persistence factor can be estimated by correlating the rainfall at individual stations with position, elevation and aspect, and assuming that the residual variance reflects the random variation. However, it is also possible to estimate the relative importance of areal pattern and random variation, by correlating the network totals between successive periods and deducing how much of the total variance could be explained by position if all the relevant factors were included. The algebra may be found elsewhere (Sutcliffe, 1966), but the standard error is reduced by the "persistence factor", which can be estimated as $\sqrt{(1 - R)}$ from the correlation coefficient R between the two independent series.

This approach is best explained by examples from two contrasting basins, the Ystwyth above Aberystwyth in Wales and the Ray above Grendon Underwood in central England. The Ystwyth basin of 200 km^2, studied by Rodda (1962), varies in elevation from 8 to 450 m with a rainfall range from 1000 to 2000 mm; the Ray basin of 20 km^2, established by the Institute of Hydrology, is relatively flat with a rainfall range from 460 to 540 mm. Two successive years of rainfall records over the two basins were compared (Fig. 3.9) and analysed; the correlation coefficient R was 0.987 for the Ystwyth compared with 0.412 for the Ray. The rainfall over the Ystwyth basin with its sharp relief is much more variable than that of the Ray; although the mean annual rainfall of the Ystwyth is over twice the Ray, the standard deviation of the individual stations is over ten times larger; much of this increased variability is due to the persistence factor because of topography. Once this is taken into account, the standard error of the Ystwyth basin rainfall is less than twice that of the Ray. This suggested that the random variation of rainfall could be much more conservative than total variation, and that consistent estimates of basin rainfall over time should be obtained from a relatively small number of gauges established over the whole period of

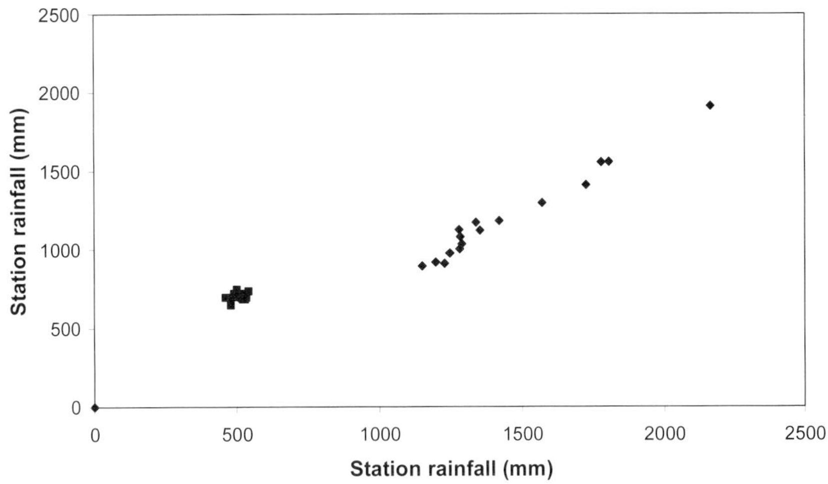

Fig. 3.9 *Correlation of annual rainfall totals: ♦Ystwyth basin, 1959 vs 1958 rainfall; and ■ Ray basin, 1965 vs 1964 rainfall.*

records rather than from a larger but varying number of gauges. This principle is discussed later in Chapter 8 in relation to Lake Victoria.

USE OF RAINFALL FOR FLOOD FORECASTING AND PREDICTION

Apart from their use in water balance appraisal, rainfall measurements are needed in flood forecasting and prediction. The distinction is that flood, or more generally flow, forecasting is carried out in real time, while flood prediction is related to probability, without reference to timing. Both activities require additional information on antecedent conditions, especially soil moisture state which is discussed in Chapter 5, but it is convenient to describe here some aspects of rainfall information required for these purposes.

Flow forecasting

Flow forecasts are needed for flood warning, to enable flood defence measures to be operated and to alert the public to the possibility of inundation, and more generally to provide the necessary information for reservoir operation. The advantage of incorporating rainfall information in flow forecasts is the gain of time and precision over forecasts based on upstream flows alone; the benefit of the time gain depends to some extent on the size of the basin, and is especially valuable in smaller basins. The type of information required is no different from that needed for water balance purposes: estimates of the variation over the basin and over time of rainfall depth. This supposes a need for autographic or continuous measurements to estimate short-term amounts, but on large basins like the Nile or Senegal daily falls are likely to be adequate.

To be useful for forecasting purposes, rainfall measurements or estimates are clearly needed in a forecasting centre as the rainfall occurs, and therefore some form of communication or data transmission is needed. The conversion of raingauges to telemetric mode is covered by Strangeways (2003), but transmission can be based on radio, telephone communication or satellite communication as in the Niger basin in West Africa.

Nevertheless, there are advantages in less traditional methods in some situations. The use of weather radar has brought benefits in visual immediacy in communications, and, after calibration of radar imagery, can provide quantitative estimates of rainfall distribution over areas where the collection and assembly of a sufficient number of direct measurements would not be practicable. The introduction of weather radar has been proposed in parts of Africa to facilitate flow forecasting for development projects, but it presupposes access to the basin which is not always possible.

In some areas it is difficult to obtain measurements from the upstream basin. In the hope that satellite imagery could be used to overcome these problems, the derivation of cold cloud duration has been used in some examples of flow forecasting systems. The network of raingauges in the Senegal basin provided historical records to calibrate the conversion of satellite estimates of cold cloud duration, the number of hours indicated by imagery below various temperature thresholds, into estimates of daily rainfall. It was found that rainfall estimates from satellite imagery could be used in simple runoff models to forecast downstream river flows with similar precision to routine rainfall measurements (Hardy *et al.*, 1989). Following this example, this technique was incorporated into a forecasting system for the Blue Nile and Atbara, where the bulk of the runoff derives from the remote part of the basin in western Ethiopia. The imagery had provided useful information during the Khartoum flooding of August 1988, when it showed that current upstream rainfall was unlikely to cause further river flooding to worsen the situation. Subsequently a flow forecasting system was developed using satellite imagery to give rainfall estimates in real time, a conceptual hydrological model to convert rainfall to river flows at key sites, and a hydraulic model to forecast levels and flows down the river system (Grijsen *et al.*, 1992).

Flood prediction

Rainfall records can provide an important contribution to flood prediction. The first requirement is a long and consistent series of daily totals from stations over the project basin, which is similar to the information required for a water balance study. However, the analysis of point rainfall data for flood prediction is usually based on the annual maximum series, which is made up of the maximum daily fall in each calendar or hydrological year of record. This series is analysed in accordance with an appropriate statistical distribution, from which the frequency with which the rainfall is likely to be exceeded can be estimated. The frequency may be expressed by its reciprocal: the return period, or the average period between years in which the rainfall is likely to exceed a given amount. Because the frequency distributions used in this analysis are also used for the analysis of flood flow records, their use is described in Chapter 10.

Similar analysis to that applied to daily rainfall series may be used for rainfall of two or more successive days, and can also be applied to series of basin rainfall, deduced in the usual way from a number of records over the basin. Where autographic or other records of shorter duration rainfall are available, annual maximum series of durations of three hours or six hours, for example, may be extracted and analysed in the same way. These analyses can be combined to derive regional relations between daily rainfall amounts and rainfall of different durations, for given frequencies or return periods; these are known as storm duration curves.

The derivation of basin rainfall from estimates of point rainfall is often based on the concept of areal reduction curves, which illustrate the reduction of average storm depth as the basin size increases; these can again be derived from relations between point rainfall and equivalent average rainfall over different areas.

An alternative to statistical analysis of rainfall, which is used for design purposes where a minimum risk of underestimation is acceptable, is a physical approach to estimation of what is known as the "probable maximum storm". In brief, the factors affecting storm amount, such as atmospheric moisture content and inflow, and the storm efficiency, are maximised from physical evidence; the resulting maximum storm is estimated. The procedure is described from examples in Chapter 10.

Before rainfall data can be used to derive an equivalent river flow, a rainfall:runoff conversion of some type is required. As described in later chapters, this depends largely on antecedent conditions in the basin, and specifically on soil moisture state; therefore further discussion of the use of rainfall estimates or statistics in flood prediction is to be found in Chapter 5 and Chapter 10.

CONCLUSION

The main use of rainfall measurements is as the major input to water balance studies, which provide the key method of checking the quality of river flow measurements and analysis, and assist in understanding the hydrological processes. However, they are also valuable for the extension of short-term flow records and identifying the critical periods for water resources purposes. In the case of regional water resources investigations, rainfall studies provide the simplest guide to variations in water supply over an area. However, the records need careful analysis, linked with an understanding of the effect of topography and aspect on rainfall distribution.

4 EVAPORATION AND TRANSPIRATION

Evaporation, the transfer of water from the basin surface to the atmosphere, is the main term facing rainfall input in the water balance equation. It is therefore an important factor in most hydrological studies, as the difference between rainfall and evaporation in its various forms largely determines the amount of water surplus in a river basin and thus the volume of river flow. It also affects the seasonal distribution of runoff; in many parts of the world the rainfall is highly seasonal, but in temperate areas where the rainfall varies less during the year, the temperature and thus the evaporation is seasonal. In terms of water resources studies, evaporation losses from storage reservoirs may be important, while irrigated crop water requirements are linked to evaporation.

However, evaporation from open water and bare soil, and the equivalent transpiration from grass or other short vegetation, and from woodland or forest, may be different from each other. Also, evaporation is not easy to measure directly, as the evaporation loss from a small pan cannot be accepted as equivalent to open water evaporation from a larger body of water. However, a number of indirect methods of estimating evaporation, which are based to a varying extent on the physical understanding of the process, are often used to estimate the different forms of evaporation.

MEASUREMENT OF DIFFERENT FORMS OF EVAPORATION

In physical terms evaporation is simply the conversion of water from a liquid form to water vapour; in this sense evaporation includes all the ways in which water at the surface of a river basin is converted into moisture in the air above the basin. However, it is generally referred to as evaporation from the surface of a water body or from bare soil, or as transpiration when it is routed through the leaves of grass or other vegetation such as woodland. A related process is evaporation from rainfall intercepted by vegetation and retained on the surface of the vegetation. There are variations in the rates of evaporation from these different surfaces, but all of them depend ultimately on an energy supply to provide the latent heat required; the evaporation rate also varies with the wind available to carry away the water vapour. These factors in turn depend in general terms on climate, so that evaporation is on the whole less variable than rainfall.

Evaporation from a large body of open water is perhaps the nearest to a standard example of the process, but is not easy to measure except as the residual from a water balance study of a lake. Measurements using evaporation pans, in which the loss of water over a known area is compared with measured rainfall, are not very satisfactory as the apparent evaporation loss varies with the dimension of the pan, and the instrument receives heat supply to the side of the pan. In arid climates the evaporating water surface is much moister than its surroundings, and thus evaporates at a higher rate than a larger body because of a so-called oasis effect. Various pans and tanks have been adopted as standard in different countries, with the United States "Class A" pan adopted as an international standard; however, empirical coefficients, which may vary with season, are required to adjust pan measurements to represent a large water body.

An early method of measuring evaporation was used in the Nile basin. Piche evaporimeters, in which a reservoir in a glass cylinder feeds a saturated disc of paper of standard dimension, were used throughout the Nile basin (Hurst, 1952) to measure the loss. This loss was then adjusted by a factor, which was taken as 0.50 in much of the basin, but which was found to vary towards 0.75 or 0.80 in more humid regions; this was clearly a rather artificial method of estimation.

Another standard definition of evaporation, known as potential transpiration, is the loss from a short grass surface which is amply supplied with moisture. Lysimeters have been designed to measure such an evaporation loss, either by continuous weighing or by adding sufficient water daily to induce drainage which is also measured. Such techniques are not easy to adapt to measuring evaporation losses from woodland, but experiments have been carried out which show these to be higher than from grass.

Although this brief discussion introduces topics relevant to later chapters, it illustrates some of the problems of direct measurement of evaporation.

ESTIMATION OF POTENTIAL TRANSPIRATION

Once potential transpiration from a moist grass surface had been proposed as a standard measure of evaporation, several approaches were developed to estimate this process from meteorological data on the assumption that the process was controlled by climate and that vegetation was simply responding to external conditions.

Thornthwaite

An empirical expression for potential transpiration was derived by Thornthwaite (1948) from records in the eastern United States of America. This formula is based on monthly mean temperatures and a heat index that is based on a complex sum of monthly temperatures throughout the year. An adjustment is made for the monthly hours of daylight corresponding to the latitude of the site. Because the relation between temperature and transpiration is derived from a specific region, this formula is not necessarily valid in other climates. A useful part of this approach is the Thornthwaite diagram (see Fig. 5.2), based on a comparison of monthly rainfall and potential transpiration; when potential transpiration exceeds rainfall, a period of deficit occurs and transpiration depends on soil moisture depletion, but when rainfall exceeds potential transpiration a period of surplus gives rise to soil moisture recharge and runoff.

Penman

The Penman (1948) approach to the estimation of open water evaporation, or alternatively of potential transpiration from a grass surface which is never short of water, was more closely based on the physical processes involved. He combined the energy balance equation, according to which net solar energy is used both in evaporation and in heating the air, and also the aerodynamic equations in which evaporation is a function of saturation deficit and wind speed, and the corresponding equation in which heat flux is a function of temperature difference between the water surface and air, and of wind speed. By assuming the equivalence of the two wind speed expressions, Penman was able to eliminate the terms involving water surface temperature.

Plate 4.1 *Meteorological station at Esquel, Patagonia. From left to right: raingauge, evaporation pan, solarimeter, anemometer, raingauge, radiometer, anemometer mast, and Stevenson screen with recording raingauge behind.*

The expression obtained is $E_O = (\Delta H + \gamma E_A)/(\Delta + \gamma)$, where E_O is open water evaporation, H is the net incoming energy, E_A is evaporation at air temperature, Δ is the gradient of saturated vapour pressure against temperature, and γ is the hygrometric constant. This expression can be solved from standard meteorological observations (Plate 4.1): air temperature and humidity, wind speed and either incoming energy or sunshine hours. The value of H may be measured directly or estimated from sunshine hours; E_A, which is the smaller term, can be estimated from an aerodynamic equation using air temperature, humidity and wind speed. Tables were compiled to assist these calculations, and nowadays computer programs are also available. Because this approach is based primarily on physical reasoning with an element of empirical observation, it can be transposed from one climate to another and it has been used in many countries as a standard method. The detailed derivation is available in many textbooks, and is not repeated here.

Penman originally introduced empirical factors, based on observation in Britain, to transform open water estimates (E_O) to transpiration from grass (E_T), but he later demonstrated that the adjustment could be made by measuring the net energy over the different surface or by altering the albedo in the estimate. A more direct approach is the Penman-Monteith equation (Monteith, 1985).

ESTIMATION OF ACTUAL EVAPORATION

An empirical formula was developed by Turc (1954), expressing actual annual basin evaporation loss as a function of annual precipitation and mean air temperature. This approach was based on comparisons of rainfall and runoff from a number of basins in many countries. Although it has been popular as a useful means of estimating potential runoff from an area, it has largely been superseded by the more direct methods based on the Penman approach.

The simplest method of estimating actual evaporation is through comparison of rainfall and potential transpiration, when the periods of soil moisture use and recharge can be deduced. Some examples of the balance between rainfall and potential transpiration are illustrated in Fig. 4.1 (a)–(c), where examples of humid, arid and mixed climates are shown. The differences between these three situations are of fundamental importance to the approach to the hydrological water balance. In fact this simple comparison epitomises the hydrological character of an area, and largely determines the complexity of the analysis. This distinction will be stressed throughout subsequent chapters, but the humid climate is the simplest to deal with, while the arid climate is the most complex.

Climate examples: Sri Lanka, Botswana, India

The humid example is from Nuwara Eliya in Sri Lanka (Fig. 4.1(a)), where average rainfall exceeds potential evaporation in almost all months; in this case actual evaporation will always be close to the potential. In the arid case of Francistown in northeast Botswana (Fig. 4.1(b)), where potential evaporation exceeds average rainfall in all months, actual evaporation will be controlled by actual rainfall rather than potential transpiration. An intermediate case is that of Bhopal, in central India (Fig. 4.1(c)), where a short period of rainfall surplus is followed by a longer period of deficit. In this case the evaporation may be assumed to equal the potential rate during the wet season; at the beginning of the dry season the evaporation will approximate to the potential while the supply is maintained by soil moisture storage. Once the soil moisture storage has been reduced by plant water use from saturation to the so-called wilting point, the evaporation will greatly reduce or cease. The difference between the water which can be held against gravity, and that which remains at the wilting point, over the soil depth from the surface to the bottom of the rooting system, is known as the root constant (Penman, 1950). This depth of water is often assumed to be transpired at the potential rate before the evaporation is reduced.

This rather simple model is illustrated by Fig. 5.2 in which monthly average rainfall and potential transpiration are compared for the Betwa basin near Bhopal. During July, the first month where rainfall exceeds potential transpiration, the surplus water recharges the soil moisture store. After the soil moisture has been recharged, the water surplus is available for runoff, which occurs between late July and the end of September. From October, the transpiration is maintained at the potential rate by soil moisture storage until the storage, which is naturally equal to the recharge in July, is used up. There follows a period of water deficit, when transpiration is much reduced. This model, which is essentially the Thornthwaite diagram, enables the actual evaporation to be estimated from the potential rate and the value of the root constant. This topic will be discussed in greater detail once the concepts of soil moisture measurement and estimation have been introduced in Chapter 5.

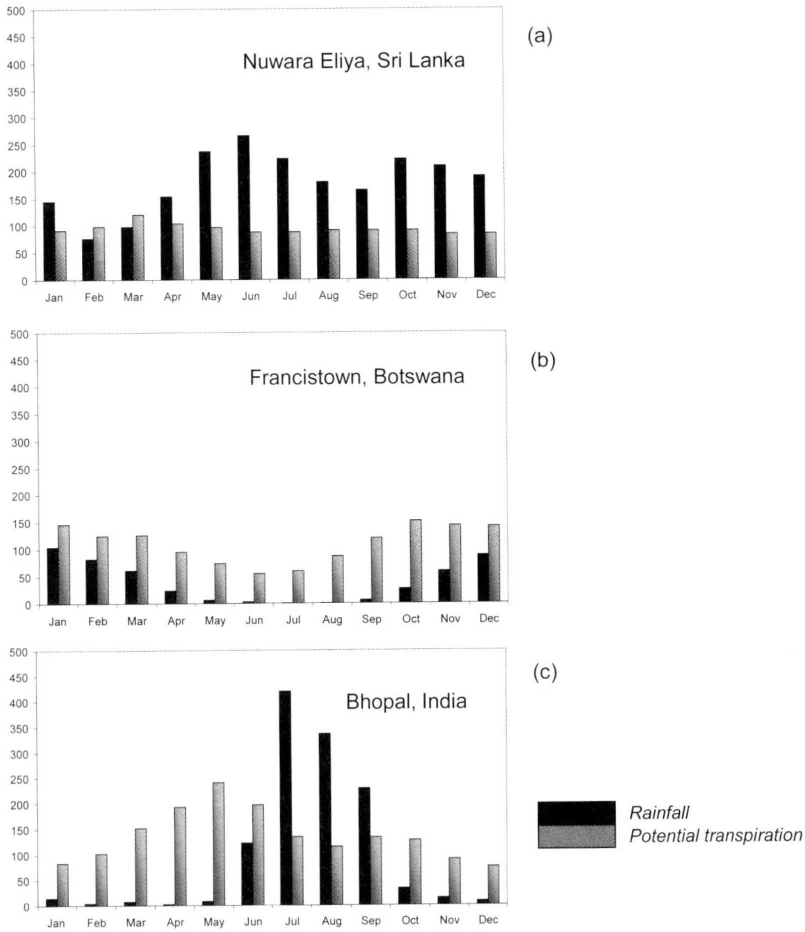

Fig. 4.1 *Comparison of monthly mean rainfall and potential transpiration (mm).*

MEASUREMENT OF OPEN WATER EVAPORATION

Botswana

In northeast Botswana there was an opportunity at Shashe reservoir near Francistown to measure open water evaporation fairly directly; the inflow was limited in most years to short periods during the summer months November–April, and spill from the reservoir was even more limited. The regime of the reservoir (Parks & Sutcliffe, 1987) is illustrated by Fig. 4.2, which shows the reservoir levels for 1984. The inflow to the reservoir was deduced from rises in reservoir level and from spill when the reservoir was full, to give a flow record which was not subject to the problems of measurement in these very ephemeral rivers. The evaporation from this open water body could be estimated from the water balance of the reservoir, taking into account inflow, spill, rainfall and abstractions, all of which were measured. Both inflow and evaporation could be estimated from a single water balance because inflow is concentrated in a small number of days.

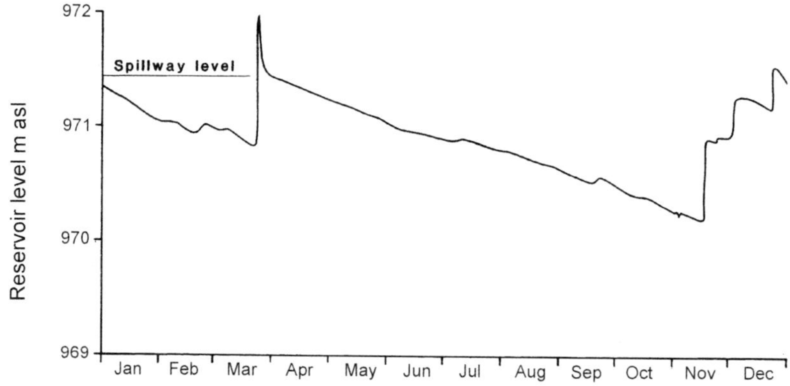

***Fig. 4.2** Daily levels of Shashe reservoir, northeast Botswana, 1984 (from Parks & Sutcliffe, 1987).*

After daily reservoir levels had been plotted to determine the periods of inflow, the periods of rise and the level changes were calculated. The inflow volumes were deduced from the level rises and the area–capacity curve. When levels exceeded the spillway level, they were converted to spill flows using an equation derived from model tests. The increases in reservoir volume and spills were tabulated separately as monthly totals. Daily rainfall totals were multiplied by current reservoir area to give rainfall volumes. The evaporation during periods of inflow and spill were estimated as a first iteration from daily rates estimated by the Penman method, and converted to volumes. The abstractions were derived from metered measurements, and the volumes during periods of inflow and spill were also deduced; because the abstractions were mainly for industrial use, they were fairly uniform.

The net evaporation was calculated from the initial and final daily reservoir levels, which were converted to area and volume. The net inflow, or total inflow less spill, was corrected for rainfall, evaporation and abstraction during the inflow period. The change of storage less corrected inflow then equals abstraction and net evaporation during the month. Seepage through the dam was known to be negligible. The net evaporation was converted to depth by dividing the volume by mean area, and the rainfall was added to give gross evaporation depth. These were tabulated and found to give an annual average of 1820 mm, which was similar to the Penman estimate. Over the 12-year period analysed, only 20 months had more than 10 days of inflow and spill, when the estimates might have been affected by measurement uncertainties. At the same time the total inflow volume was estimated for a 15-year period, with an annual average of 91 $m^3 \times 10^6$.

The results of this water balance analysis for 1984 are illustrated in Fig. 4.3 where monthly inflow volumes and evaporation depths are shown. They demonstrate the importance of reservoir evaporation in this semiarid example. Whereas net evaporation from a reservoir can almost be neglected in countries like Britain, where most reservoirs are built in upland areas where rainfall exceeds evaporation, the evaporation from reservoirs in Botswana is a most important component in the analysis and in fact is several times the net yield of the reservoir itself.

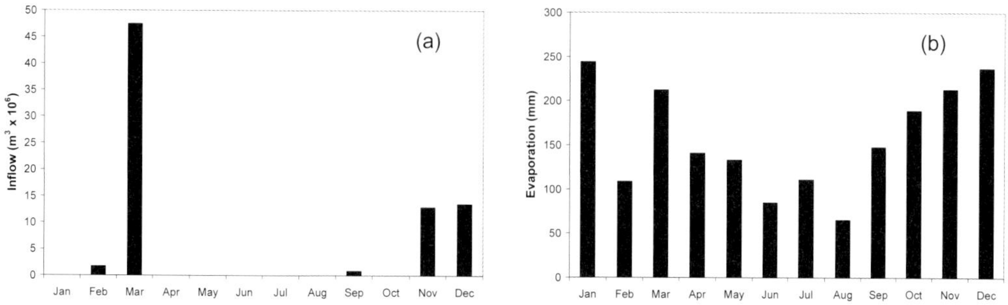

Fig. 4.3 *Water balance components for Shashe reservoir, 1984: (a) inflow ($m^3 \times 10^6$); (b) evaporation (mm).*

This example illustrates the fact that direct measurements of open water evaporation are complex, even when conditions are extremely favourable, as in this case of a reservoir with measured abstraction in an arid climate where periods of inflow are brief. Similar calculations can be made in other climates, but the result may depend on measurement of the differences between large numbers; the various components are each measured with given errors, but the differences between inflow and outflow components are estimated with greatly enhanced error where the components are of similar magnitude.

ESTIMATION OF WATER USE BY VEGETATION

The estimation of crop water use depends on the conversion of potential transpiration from grass to the potential transpiration from crops in various stages of growth and maturity, for which tables of crop coefficients (Doorenbos & Pruitt, 1977) are available. In general crops transpire at a rate below the potential when they are growing and not yet covering the ground surface, then at the potential rate when they are fully grown, and finally at a reduced rate when they reach maturity and are about to be harvested. The net irrigation requirement can be deduced from the gross requirement by allowing for rainfall during the periods of growth and maturity. Perennial crops, on the other hand, are likely to transpire at near the potential rate if water is available and thus their water needs will depend largely on climate.

One aspect of evaporation from vegetation is the effect of interception of rainfall by the leaves and stems of vegetation, especially woodland. Much research on this topic (e.g. Gash, 1979) has shown that the depth or fraction of rainfall intercepted depends not only on the type of vegetation but also on the time distribution of rainfall. For instance, the annual interception by broad-leaved woodland with rainfall distributed throughout the year, will exceed that from acacia-type vegetation in rainfall limited to intense storms in a short rainfall season. Much of the intercepted water will be evaporated from the leaf surface, and can be regarded as part of the water use of vegetation; the energy use may be increased by lateral flow or advection.

Water use in Bahrain

During a study of the water resources of Bahrain in 1967, the water balance of the island was found to depend on the balance between the groundwater inflow, the sole source of supply (which was derived from recharge in Saudi Arabia and flowed

Plate 4.2 *Date palm plantation in Bahrain.*

through the deep limestone aquifer under the sea), and groundwater abstraction for domestic and industrial use and for crop water use. It was necessary to make an estimate of water abstraction on the island which could be compared with the estimated inflow.

By 1967, the number of drilled wells had reached almost 800, and the measurement of even a sample number would have been a large task. The groundwater abstraction for urban and industrial water supply was measured, but the most important component was the abstraction from springs and wells for irrigation of date palm plantations. Because these irrigated areas were clearly shown on maps and air photography, the simplest way to estimate the abstraction was through the transpiration from these areas. It was possible to confirm the areas and estimate the densities of these plantations by field reconnaissance coupled with study of air photography. The water use was estimated from the product of the effective area and the transpiration rate. From the density of the date palms (Plate 4.2) the transpiration rate was assumed to correspond with the open water evaporation rate, which was estimated by the Penman method from meteorological records as 2580 mm per year; the annual rainfall of below 75 mm could be neglected in this case. Allowance was made for irrigation distribution losses, and leaching requirements to ensure drainage, at 25–50% of evaporation. This indirect estimate provided the major item in the water balance; with municipal and industrial use the annual abstraction totalled 122–147 $m^3 \times 10^6$. The use of this estimate is described in Chapter 6.

Abu Dhabi

In another case, in a rapid study of the water resources of Abu Dhabi, a major source of recharge to the main aquifer was runoff from the mountains of the Jebel Akhdar from which a number of streams emerged on to alluvial fans near the border with Oman. No measurements of these streams had been made, but it was possible after visiting the region to estimate their flows (see also Chapter 6) using the areas of perennial date palms and other crops irrigated near villages along the streams. Some of the agricultural areas relied on base flows diverted into channels by temporary dams on the main streams. Others relied on flood flows which recharged through fans into the alluvial aquifers; this subsurface flow was diverted from relatively narrow river valleys downstream, where flows were concentrated, by means of *aflaj* or underground tunnels similar to the *qanats* of Iran. The combined flows could be deduced from evaporation from the areas of perennial irrigation near these two types of villages, which were termed surface water and groundwater villages.

MEASUREMENT OF WETLAND EVAPORATION

Large wetlands, with a mixture of open water and swamp vegetation either floating or anchored in saturated soil, provide a surface with some similarity to open water; there is no complication caused by the transpiration decreasing as a result of soil moisture deficit. However, it is not easy to measure evaporation and transpiration directly.

The Sudd, the area of swamps of the Bahr el Jebel in the southern Sudan, where about half the inflows are lost to the White Nile outflow due to evaporation, provides an example of a large wetland in an area of moderate rainfall; the existence of the wetland depends on the inflow fed by Lake Victoria and the other East African lakes.

Direct measurement

At a time when the estimation of evaporation from meteorological data had not been developed, an attempt was made to measure evaporation (Hurst & Phillips, 1938) near the centre of the swamps by establishing a tank 10 m square and 2.5 m deep. The tank was sunk in the swamp and planted with papyrus; the water in the tank was kept at the same level as the swamp, and the loss of water was measured regularly and adjusted for rainfall. It was noted that losses were higher on rainy days, which was attributed to interception. The observations from 1926 to 1932 gave an annual mean of 1318 mm, but the annual totals decreased over this period. It was noted that the papyrus within the tank was less luxuriant than that outside, and it was found difficult to maintain typical growth.

Moreover, the evaporation losses deduced from the average area of flooding, compared with inflows, rainfall and outflows, were insufficient to explain the losses in the Sudd. Hurst & Phillips (1938) suggested that evaporation must have been underestimated by about 30% in order to explain the balance. In later studies, Migahid (1948, 1952) carried out a botanical investigation of the area and improved the evaporation tanks, with growth in the tank as good as that outside. Measurements in 1947–1948 gave annual results approaching 2400 mm, which were similar to losses from a tank full of water in an open lagoon. The interception by vegetation was measured as 45% of rainfall. Penman (1963) commented that the diurnal weather cycle, with relative humidity decreasing to a minimum during the day as the wind speed reaches a maximum, would lead to extreme evaporation and to transpiration from papyrus similar to evaporation from an open lagoon.

Indirect estimation

These experiments confirm that the direct measurement of wetland losses is difficult. In many ways it is easier to establish evaporation rates as a residual of the water balance. In an investigation of the Sudd to deduce the potential effect of the Jonglei Canal on areas of flooding, Sutcliffe & Parks (1982, 1987) developed a simple hydrological model. Using monthly river inflows and outflows for the period 1905–1980, rainfall measurements (average 870 mm) and swamp evaporation taken as open water evaporation (2150 mm) estimated by the Penman method, they estimated monthly areas of flooding. They took account of infiltration into newly flooded areas, and used a linear relation between areas and volumes of flooding to convert cumulative differences between inflow and outflow volumes to flooded areas. The series of flooded areas covering the period of flow records were compared with a number of estimates of these areas, taken from air photography and satellite imagery or from vegetation maps compiled at different dates. These comparisons (see also Chapter 8) showed that the modelling was realistic and therefore that the Penman open water evaporation was a reasonable estimate of evaporation from the wetland.

ESTIMATION OF NET RAINFALL

In the case of open water evaporation from a reservoir or evaporation from a wetland, the net rainfall can be deduced simply from the difference between rainfall and evaporation; in fact annual rainfall was less than evaporation in the case of the Sudd, so the net rainfall was negative. When the actual evaporation reduces below the potential transpiration rate following the development of a moisture deficit, the derivation of actual evaporation from the potential rate must take account of the soil moisture state. Further discussion of actual rather than potential evaporation follows the discussion of soil moisture storage in the next chapter.

CONCLUSION

Evaporation takes a variety of routes, directly from open water or soil, or from intercepted water, or alternatively as transpiration through vegetation, but in general the process is controlled by climate and may be estimated from meteorological records. The process is reasonably conservative in time and space, and fewer measurements or estimates are required than for rainfall. It is possible, but not usually straightforward, to measure evaporation directly in some cases, but it is often deduced as the residual of the water balance.

5 SOIL MOISTURE STORAGE

Soil moisture storage has a physically important role in the hydrological cycle, and it has a vital influence on the amount of rainfall which becomes runoff and groundwater recharge. The problems of measuring soil moisture volume and the equivalent depth of water contained in the soil reservoir have led to various indirect methods of measurement. The role of soil moisture storage in maintaining vegetation or crop growth during dry periods is important and it plays a key part in controlling runoff in a monsoon climate; soil moisture investigations played an important part in an investigation of basin water balance and groundwater recharge in central India, which is described here and later. The role of soil moisture state in converting storm magnitude to flood volume is illustrated by evidence drawn from British studies.

PHYSICAL CHARACTERISTICS OF SOIL MOISTURE STORAGE

Soil moisture storage is a key factor in the water balance of a basin, as it provides a short-term reservoir maintaining the continuous outputs to basin evaporation and transpiration from the discontinuous rainfall inputs. The volume, or equivalent depth, of soil moisture storage between the ground surface and the lower limit of the rooting zone is an important indicator of the state of the basin at any given time. In simple terms, as illustrated in Fig. 5.1, soil moisture varies between field capacity, i.e. the maximum quantity of water which can be held in the soil against gravity, and wilting point, below which vegetation cannot extract further water. Both water content limits may be expressed as percentages of the total soil volume; the water content held at these two limits depends largely on soil characteristics. The total water volume within a soil profile which is stored and available to vegetation between these two limits also depends on the rooting depth of the specific vegetation; this volume is known as the root constant. The soil moisture deficit, or the depth of water required at a given time to bring the soil moisture reservoir to field capacity, largely controls the runoff response of the basin to a given depth of rainfall. In general, the rainfall infiltrates into the soil layer and brings the storage to field capacity before significant runoff occurs, though surface runoff can occur over short periods if the intensity of the storm exceeds the maximum infiltration rate of the soil surface. Any rainfall input when the soil is at field capacity soon drains, i.e. forms recharge if the sub-surface is permeable, or, on impermeable strata forms runoff to streams.

Thus the soil moisture state to a large extent determines how much of the rainfall input is immediately released as river flow. This is often expressed as a runoff coefficient, or the fraction of rainfall which runs off, though this is not a physically accurate concept; runoff is more truly a residual of the rainfall–recharge process. Soil moisture storage may be measured in various ways, or may be estimated by the indirect method of soil moisture modelling. Examples of both soil moisture measurement and modelling are described.

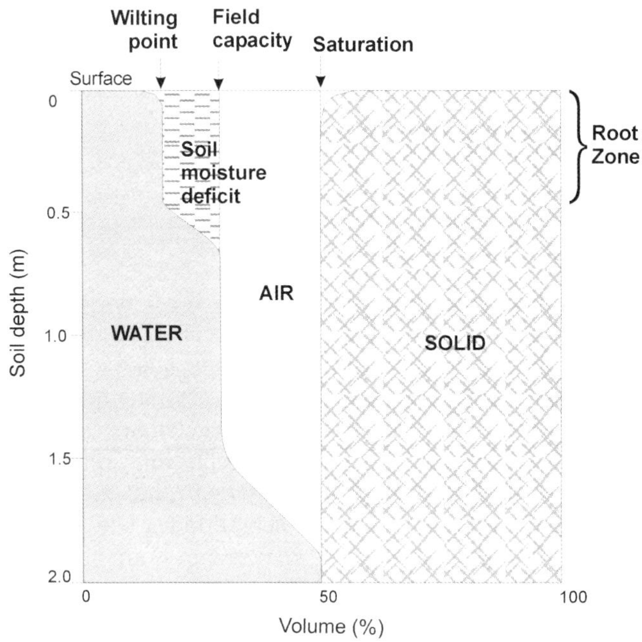

Fig. 5.1 *Soil moisture: field capacity, wilting point, soil moisture deficit.*

SOIL MOISTURE MEASUREMENT

Early observations of soil characteristics in wet conditions were included in the flood plain survey of the Bahr el Jebel or White Nile in the southern Sudan during 1951–1952. Soil pits were dug to a depth of 1.8 m at kilometre intervals along cross-sections. Samples were collected to supplement visual observations and notes of the soil characteristics, but in these field conditions soil moisture content could only be estimated by observing the depth at which standing water occurred in the pits. Chemical and mechanical characteristics of soil samples at 30 cm vertical intervals were derived by analysis, and a soil map (Jonglei Investigation Team, 1954) revealed the pattern of soil deposition across the flood plain, with the coarser material deposited near the river and finer material carried further.

In order to measure soil moisture directly networks of sites would need to be sampled regularly, by digging or soil augering, with the moisture content of samples determined by drying in ovens and the weight loss measured (ISO, 1993, 1997a). In the absence of such facilities, a subjective test of soil moisture was devised by the soil specialist of the team; samples were thrown up to the office ceiling and observed whether they stuck or fell! Various more practical devices, including tensiometers and electric resistance gypsum blocks, were developed to measure soil moisture *in situ*, and these were used in basin experiments, e.g. Pereira, 1962. Gardner *et al.* (2001) provide an overview of current soil water content measurement methods.

Neutron probe

The regular large-scale *in situ* measurement of soil moisture was made practicable when the neutron probe was developed for the purpose (Bell & McCulloch, 1966; Bell,

1976; Gardner *et al.*, 2001). This is used to measure the vertical moisture content profile by lowering a probe with a fast neutron source down an aluminium tube and counting the back-scattering of neutrons near the source; this reflects the moisture content of a sphere about 50 cm in diameter around the source and can be calibrated using local soil samples.

This technique was used to monitor the moisture content at a network of sites in the experimental basins of the Institute of Hydrology on the Ray at Grendon Underwood in central England and the upper Severn and Wye basins at Plynlimon in Wales. The Ray experimental basin was set up to monitor the water balance of the basin in order to develop modelling techniques (Nash, 1960; Mandeville *et al.*, 1970), while the Wye and Severn basins were monitored to observe the differences in behaviour of adjacent basins with contrasting land cover of forest and grass (Kirby *et al.*, 1991). It was therefore important to measure or estimate the soil moisture at regular intervals at the same time as other components of the water balance were measured. After a reasonable set of soil moisture data had been measured, it was found possible to calibrate a soil moisture deficit model for the Plynlimon basins which gave reasonable estimates of soil moisture conditions.

Betwa study, central India

An investigation of groundwater conditions and recharge in a hard-rock basin in the Betwa basin in central India was undertaken from 1976 to 1980. The study was a joint research project between the Indian Central Groundwater Board and the British Geological Survey, and the Institute of Hydrology was responsible for the surface water and soil moisture aspects of the study.

The climate of this basin is highly seasonal, with rainfall concentrated in the monsoon between June and September, with over 90% of the annual total occurring in this period. An analysis of the water balance of the basin was carried out at an initial stage of the study (Sutcliffe *et al.*, 1981). Rainfall records for the whole Betwa basin were combined, as described in Chapter 3, to give a monthly series for the years 1926–1975. Meteorological records were used to estimate open water evaporation and potential transpiration by the Penman method. Average monthly totals of rainfall and potential transpiration are compared in Fig. 5.2 to illustrate the seasonal regime.

The annual water balance may be simplified by considering the seasonal cycle as a single period of water surplus during the monsoon and a single period of deficit during the rest of the year. During the monsoon period, monthly rainfall exceeds potential evaporation for two to four successive months, while potential transpiration exceeds rainfall during the rest of the year. The period of soil moisture recharge is indicated in Fig. 5.2, and is followed by a period of water surplus or runoff; after the monsoon ends soil moisture is used for transpiration until the soil moisture storage is depleted. This is followed by a period of water deficit after the storage has been used; the soil moisture recharge equals the subsequent soil moisture use. In the Betwa basin, the soil moisture storage is full at the end of the monsoon, and always reaches wilting point before the end of the dry season; thus the annual recharge brings the storage from wilting point to field capacity throughout the root zone. This difference is the "root constant". The soil moisture recharge should be reasonably uniform from year to year in the absence of land-use change and may be considered as a first charge on the net rainfall, the surplus

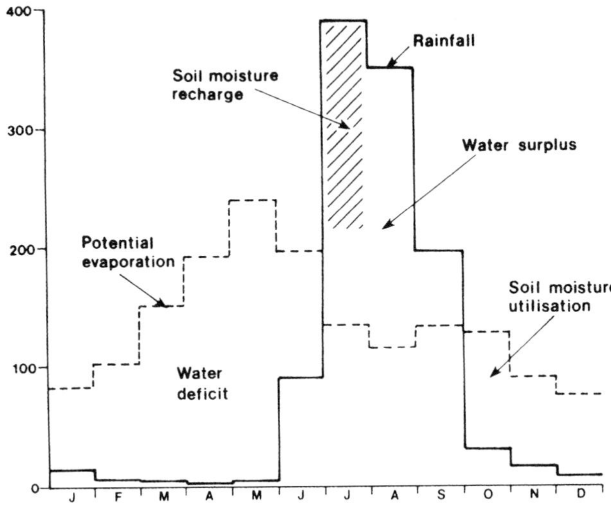

Fig. 5.2 Betwa basin: net rainfall, soil moisture recharge and runoff, mm (after Sutcliffe et al., 1981).

of rainfall over evaporation. In this seasonal climate it is possible to assume that basin evaporation continues at the potential transpiration rate until the soil moisture deficit reaches wilting point. The root constant, or the available soil moisture reservoir, can then be estimated by comparing annual net rainfall with annual runoff.

This net rainfall, or gross seasonal rainfall, was calculated by subtracting monthly potential transpiration from monthly rainfall, and by adding the surplus monthly figures to give the net rainfall for the year. If one accepts that soil moisture recharge is a fixed charge on the net rainfall, it is possible to deduce the annual recharge by comparing seasonal net rainfall with annual runoff. The intercept on the horizontal axis provides an estimate of the annual soil moisture recharge. In the case of the Betwa basin of 20 600 km^2, annual total inflows to reservoirs at the lower end of the basin over the same 50-year period were converted to runoff depths and compared with net rainfall in Fig. 5.3. The soil moisture recharge is shown to be just below 200 mm, while the mean runoff is 350 mm over the basin. Gross and net rainfall may be compared to runoff in mm to illustrate the increasing variability of the water balance.

It was evident during this early part of the study that the soil moisture storage formed a crucial part in the water balance of the whole basin. Apart from maintaining crop growth after the monsoon period, it was also a key to the conversion of rainfall to runoff. In fact the measured values of soil moisture in part of the Betwa basin may be used to monitor the partition of rainfall during the monsoon season between transpiration, soil moisture recharge, groundwater recharge and finally runoff, as described in Chapter 8. The groundwater recharge was found to be small and equal to the dry season river baseflow.

Soil moisture network

Once the key role of the soil moisture storage was recognised, a network of soil moisture observation points was established (Hodnett & Bell, 1986) in the very flat plain across the centre of the basin (Plate 5.1). Deccan Trap basalts are found over

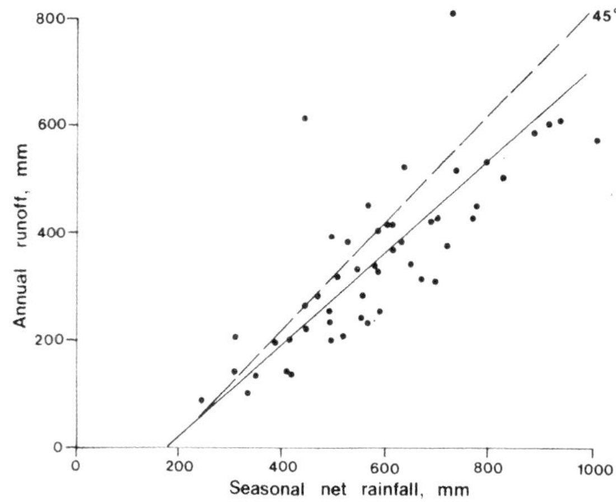

	Mean (mm)	Std (mm)	CV %
Gross rainfall	1138	211	0.185
Net rainfall	583	170	0.292
Runoff	351	157	0.447

Fig. 5.3 *Betwa basin: seasonal net rainfall and runoff (mm), and annual totals (after Sutcliffe et al., 1981).*

Plate 5.1 *Neutron probe soil moisture measurement in the Nion basin.*

most of the basin infilling the pre-existing landscape of Vindhyan sandstones. Over the centre of the basin, the soils developed from the basalts are dark swelling silty clays, known locally as black cotton soils. On the fringe of the plain, the cotton soils are from 0.5 to 2.0 m deep, and directly overlie the weathered basalt. Over the central part of the basin, the black cotton soils are about 2 m deep and merge down into the silty yellow clay from which they developed. The combined depth of the black cotton soil and

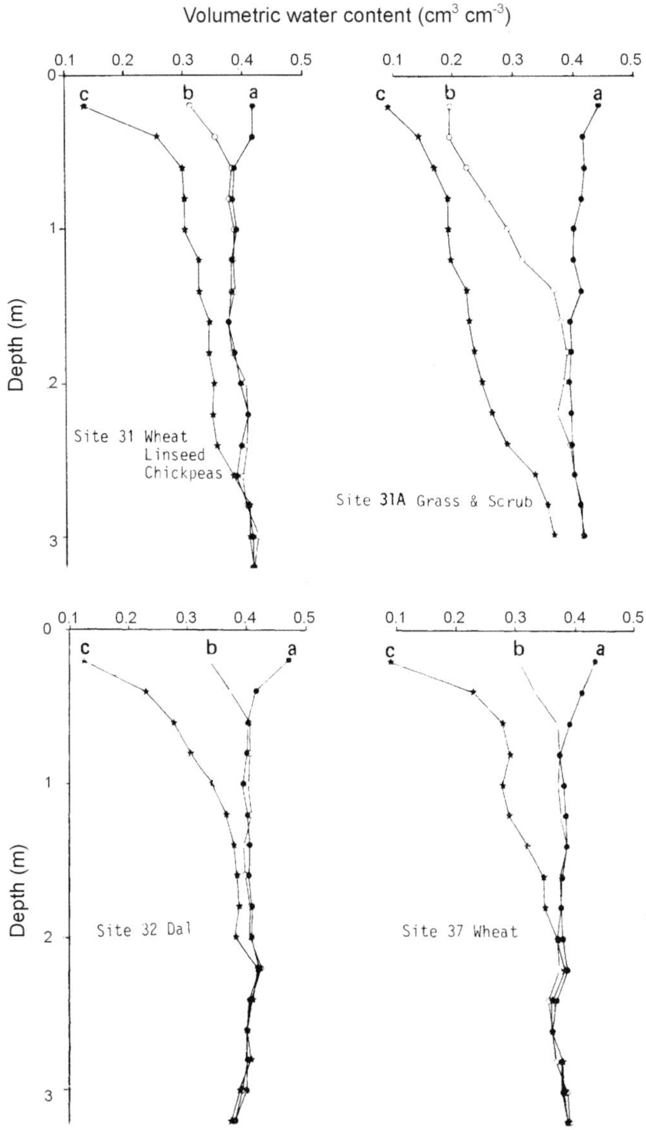

Fig 5.4 *Betwa basin: soil moisture profiles (from Hodnett & Bell, 1986).*

yellow clay overlying the weathered basalt varies from 5 m to over 20 m. The soils are remarkably uniform and are stone-free; they shrink and crack extensively during the dry season. About 85% of the area was cultivated, with wheat, pulses and linseed the main crops grown in the post-monsoon season. The local mean annual rainfall of about 1200 mm is slightly higher than the basin average.

At 24 sites along a 120-km transect across the basin, soil moisture was measured to sample the different soil depths and typical crop types of the area. Neutron probes were used in tubes which were up to 3.2 m deep, and measurements were made at weekly intervals over a complete year, starting at the end of the wet season. The moisture content was monitored at each site at vertical intervals of 0.2 m and profiles

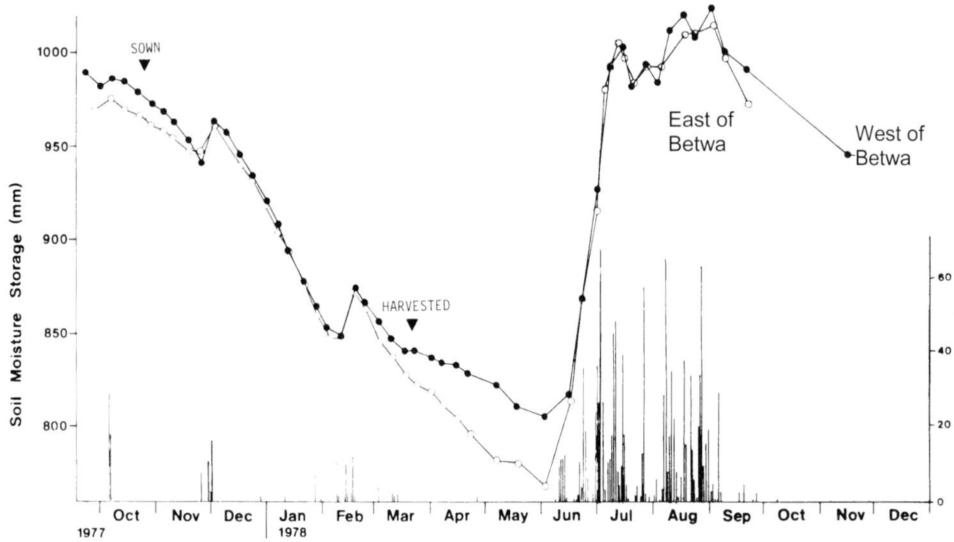

Fig. 5.5 *Betwa basin: soil moisture storage and daily rainfall, September 1977–December 1978 (from Hodnett & Bell, 1986).*

were plotted to show the total moisture content. The successive drying of the soil profile could be observed during the course of the post-monsoon season when the crop water supply was derived entirely from the soil moisture storage. Examples of three states are illustrated in Fig. 5.4 for four sites in the area of deep soils with: (a) the wettest conditions, (b) those when the water table had fallen below the measured profile, and (c) the driest conditions shown at three cropped sites with one under grass and scrub (Hodnett & Bell, 1986).

The wet season profiles at different sites in the region of deep soils were extremely similar, with very little variation in total moisture content. Under cropped sites, the soil moisture content was reduced during the season of crop growth after the monsoon over a depth of about 2 m, and the mean decrease in moisture content was 230 mm. The depletion below the site with a permanent cover of scrub and grass was 550 mm and extended below the maximum depth of measurement (3 m). This detailed measurement was successful in monitoring the soil moisture reservoir which maintains annual crop growth after the monsoon in this region of highly seasonal rainfall supply. The soil moisture storage from September 1977 to December 1978, including the crop period from October 1977 to March 1978, is compared with daily rainfall in Fig. 5.5.

Supplementary measurements of soil moisture content by neutron probe, and of soil water potential by tensiometers, were carried out in 1978 and 1979. Profiles of soil moisture content and potential, together with estimates of hydraulic conductivity derived from gradients of hydraulic potential, enabled the downward flux to be estimated.

APPLICATION TO OTHER AREAS

The simple approach to seasonal water balance derived from the Betwa basin is also appropriate to areas with similar climate, especially seasonal rainfall distribution. The approach was applied (Sutcliffe & Piper, 1986) to the problem of assessing runoff in

Guinea, where the West African rainfall is highly seasonal, with rainfall between June and September but rather higher in total than in the Betwa basin. The annual rainfall varies from over 4000 mm along parts of the coast to about 1300 mm in the northeast of the country where the upper Niger flows into Mali. Much of the basin is covered with deciduous woodland, which suggests that similar processes occur to the Betwa basin, and the same assumption could be made that a single wet season is followed by a single dry season during which the soil moisture store is drawn down to wilting point. Estimates of annual river flows were required for reviews of potential hydroelectric power resources at a number of sites over the country, and long-term rainfall records were available for analysis at a number of sites, supported by flow records for different periods.

Monthly rainfall series at all stations were converted to net rainfall series in the same way as in the Betwa example, by subtracting potential transpiration and summing the seasonal surplus. Net rainfall series for individual basins were deduced from the available rainfall stations, and annual totals of net rainfall were compared with measured annual river flows. As Fig. 5.6 shows, this comparison was reasonable where an adequate number of rainfall stations was available; the relationship has a slope of about 45° and an intercept of about 300 mm. Another comparison (Fig. 8.23) showed that the mean annual runoff depth at all stations was related to the net rainfall of their basins with a loss, corresponding to an annual soil moisture recharge, of about 300 mm. This suggested not only that annual runoff series at stations with short periods of record could simply be extended by means of annual net rainfall series by subtracting the same annual loss, but also that river flows at ungauged sites could be deduced from mean basin net rainfall in the same way.

Isohyetal maps of gross and net rainfall, Fig. 5.7(a) and (b), were compiled from the series for individual stations, and mean annual net rainfall for basins with reasonably long flow records was compared with mean runoff. The mean runoff showed good correlation with mean net basin rainfall, and corresponded with mean

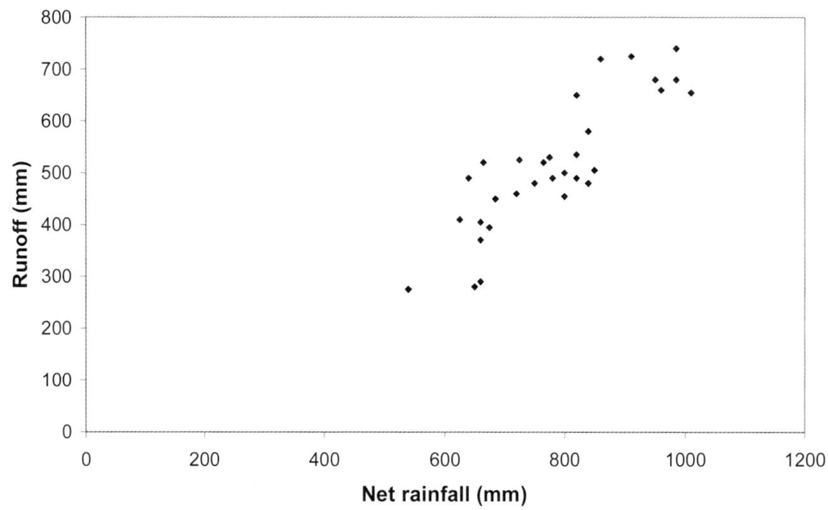

***Fig. 5.6** Guinea: Niger at Tiguibery, annual net rainfall and runoff (after Sutcliffe & Piper, 1986).*

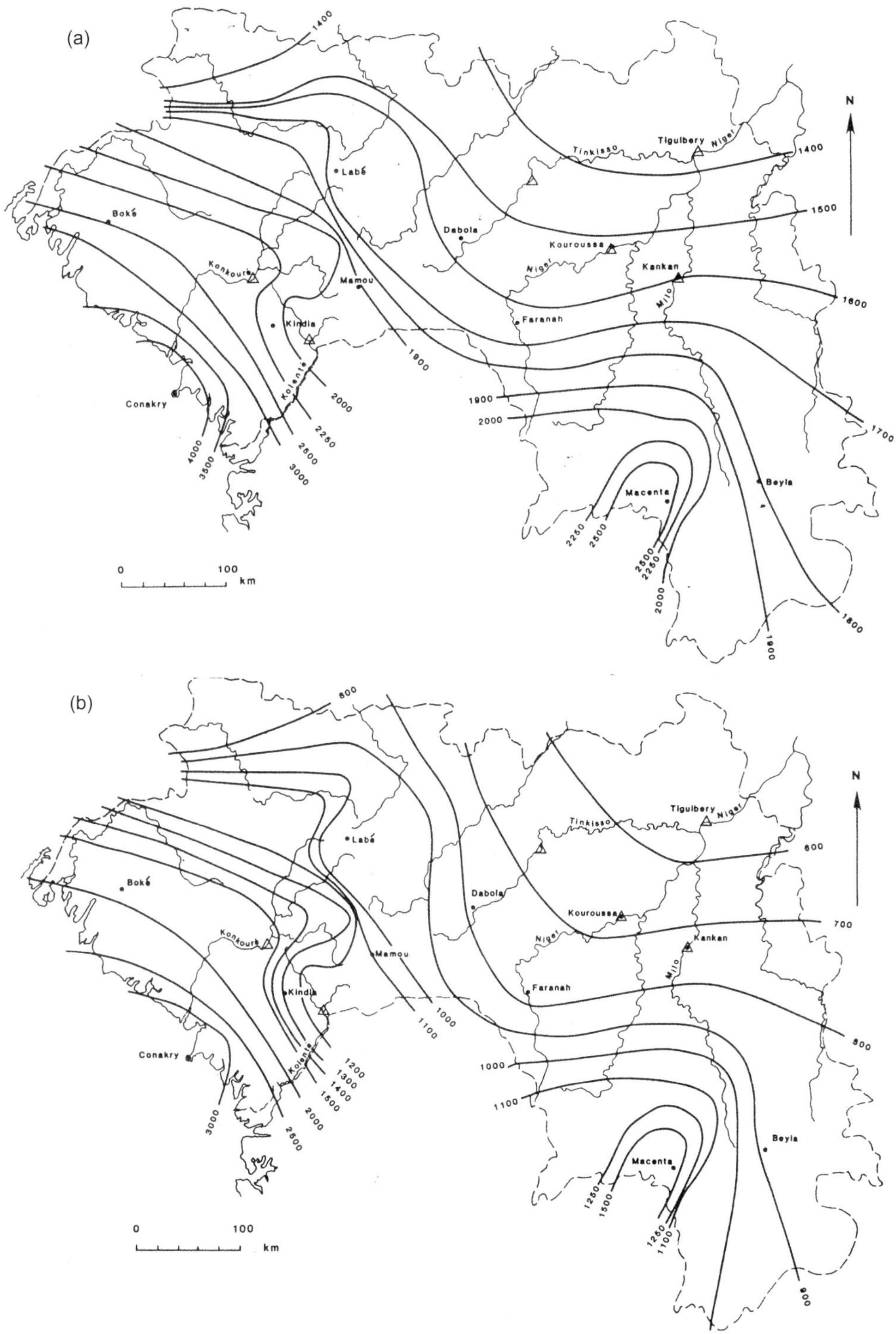

Fig. 5.7 Guinea: (a) gross annual rainfall (mm) 1951–1980; (b) net annual rainfall (mm); Δ river gauging station (from Sutcliffe & Piper, 1986).

basin net rainfall less 300 mm. Thus annual runoff series could be estimated with reasonable precision, even at ungauged sites, by deriving seasonal net rainfall series with an allowance of 300 mm for annual soil moisture recharge.

It is interesting to note that when the same approach was applied to Togo and Benin, the corresponding graphs linking annual runoff depths to annual net rainfall had slopes of 45° only in the north where a single rainfall regime is found, and not further south where two rainfall seasons occur.

Relevance to irrigation requirements

The importance to irrigation of soil moisture storage, and in particular the root constant, is that it largely determines the frequency with which irrigation applications are necessary. The rate of crop water use is related to net evaporation, or the potential transpiration and crop factor less rainfall; however, the time interval between water applications is also dependent on the volume of water on which the crop can draw, which is related to the root constant.

USE OF SOIL MOISTURE DEFICIT IN FLOOD STUDIES

During the investigations leading to the *Flood Studies Report* (NERC, 1975), a statistical comparison was made of flood statistics, in particular the mean annual flood, at over 500 gauged sites throughout the British Isles, with basin characteristics which could be measured from maps or derived from climate statistics. A major objective of this study was to develop a method of estimating the flood potential of ungauged basins. To this end a search was made for an index of flood-producing rainfall which should be more closely related to floods than indices used in earlier studies, like the mean annual rainfall or the daily rainfall of a given return period. It was thought desirable to take account not only of short-period rainfall but also of the antecedent conditions affecting the portion of the rainfall which runs off. In general annual rainfall is over 600 mm, distributed throughout the year; on the other hand evaporation, and therefore soil moisture deficit, is highly seasonal. It had previously been found that in British conditions (Mandeville *et al.*, 1970) the runoff volume could be predicted by subtracting the antecedent soil moisture deficit from the gross rainfall to give rainfall excess. Because the soil moisture deficit is the amount of water needed to restore the soil to field capacity, it was hoped that the basin rainfall excess would provide a realistic index of regional flood potential.

Although worldwide practice had been largely based on the premise that flood-producing rainfall depends on a rainfall intensity which exceeds soil infiltration rates, and thus gives rise to surface runoff, it was felt that climate conditions in the British Isles very rarely produced surface runoff in this way, as rainfall intensities were normally relatively modest. During lectures to an international course in Britain, it was interesting to note that almost all those who claimed to have seen surface runoff were overseas students!

During the development of unit hydrograph techniques of modelling flood events, records were compiled of storm characteristics and basin response in the form of immediate runoff volumes. These records were used in a qualitative statistical test (NERC, 1975, I.6.5.5) to see whether the water balance model, the infiltration model, or a contributing area model, best explained the relation between storm and response.

In the water balance model the loss between storm volume and immediate runoff was assumed to equal the soil moisture deficit; in the infiltration model the loss was related to storm intensity; with the contributing area model the runoff coefficient would vary with soil type and soil moisture deficit. By expressing these different assumptions algebraically and rearranging, it was deduced that the first concept implied that the loss, or difference between storm depth and immediate runoff, would vary with soil moisture deficit; in the second case the loss would be related to storm duration; in the third the loss would be related to storm depth and soil moisture deficit.

Nearly 1500 actual events were used to test these approaches; records included storm depth and duration, runoff volume, soil type and a catchment wetness index, which combined antecedent soil moisture deficit and recent rainfall. The loss was found to be related to rainfall depth, soil type and antecedent conditions; the duration of the storm was not significant. It was deduced that soil moisture state rather than rainfall intensity had more influence on the proportion of rainfall which contributed to immediate runoff. Therefore, in the unit hydrograph modelling, the runoff coefficient was related to a wetness index based on soil type, the soil moisture deficit, and recent rainfall.

Development of net rainfall index

It was also decided that a net rainfall index, combining short-term rainfall and soil moisture deficit, should be developed as a basin characteristic for comparison with flood statistics in a countrywide statistical study. In the same way that an index of gross rainfall, of say 5-year return period, should be more relevant to flood statistics than mean annual rainfall, it was hoped that an index of net rainfall would be even more useful. The aim was to produce a map of daily rainfall less soil moisture deficit, of 5-year return period.

At the time, daily rainfall data for about 400 stations were being analysed by the UK Meteorological Office to provide rainfall frequency statistics over the country. Soil moisture statistics had also been calculated (Grindley, 1970) from meteorological records using a continuous water balance model based on Penman (1950), estimating soil moisture storage from rainfall and potential evaporation data; a typical basin was assumed with fixed proportions of short-rooted, deep-rooted and riparian vegetation with their appropriate root constants (75 mm, 200 mm, infinite). Evaporation was assumed to occur at the potential rate until the soil moisture deficit reached the root constant, and when the deficit was zero the rainfall excess was assumed to become runoff. Statistics of estimated soil moisture state were available for 180 stations covering periods up to 30 years.

It would have been possible to analyse records of rainfall excess by subtracting the soil moisture deficit on the date of each storm, but this would have limited the number and period of records available. The alternative was to analyse rainfall and deficit records separately and to derive the statistical distribution of their difference. It was found that the two variables could be rendered independent by considering storm events larger than a threshold exceeded on average 10 times a year, and by dividing the year into seasons. This meant that it was possible to deduce the distribution of the net rainfall from the two distributions of the gross rainfall and the deficit. It was also found that samples of daily rainfall above a threshold could be treated as drawn from an exponential distribution. These two findings greatly simplified the derivation.

It was shown (Beran & Sutcliffe, 1972) that if the distribution of rainfall is exponential, and the soil moisture deficit (SMD) is independent of rainfall amount, then the net rainfall or the difference between the two also has an exponential distribution with the same slope. It follows from the geometry of the exponential curve that the difference between gross and net rainfall of the same probability is a constant which is independent of probability. This theory was tested by examining the difference between gross and net rainfall, or the effective mean SMD, at different frequencies at a large number of stations; it was concluded that the value of the mean SMD at a return period of 5 years was valid from the mean annual event up to the 100-year event. This amount can be described as a weighted mean soil moisture deficit for the given location and season. Seasonal maps of mean soil moisture deficit could be prepared and used together with rainfall statistics to provide an index of flood-producing rainfall. A map of effective mean soil moisture deficit for the British Isles was derived from analysis of some 200 stations and published as part of the *Flood Studies Report* (NERC, 1975). The geographical distribution of this mean soil moisture deficit was clear; it increased from less than 2 mm in parts of northwest Scotland, central Wales and western Ireland to 10–20 mm in southeast England. This map was intended for use in combination with maps of rainfall of 5-year return period to give an index of net rainfall. Maps of rainfall excess of 5-year return period were later derived and published for the British Isles for use in flood estimation (Farquharson *et al.*, 1975).

Thus regional use of the concept of soil moisture deficit, viewed as a function of climate, was able to provide an additional tool to add to the normal analysis of rainfall statistics.

CONCLUSION

Several different examples of the use of measured or estimated soil moisture deficits to describe the conversion of rainfall to runoff illustrate the important role which soil moisture plays in the hydrological cycle. It is not an exaggeration to suggest that it is the key to understanding the difference between potential transpiration and actual evaporation, and thus the role of the basin in converting rainfall to soil or groundwater storage and river flow. It could be argued that the fairly recent ability to measure and thus to model soil moisture state was an essential condition to the ability to model the hydrological cycle with a degree of realism.

6 GROUNDWATER RECHARGE

Groundwater studies have traditionally been linked with geology, however, hydrogeology has become a subject in its own right, with its own techniques of analysis, but based initially on geological investigations. This chapter does not aim to cover the whole field, which would require a book on its own; various hydrogeological textbooks are available (e.g. Price, 1996; Rushton, 2003). It concentrates on the links between surface water and groundwater and how information from one field can help in understanding processes in the other. The example of groundwater recharge illustrates ways in which the study of the surface water balance can assist hydrogeological analysis, while the study of the base flow component of runoff can give information on groundwater storage. With the help of groundwater analysis methods, examples illustrate the way in which measurements of different components of the groundwater process make a more precise analysis possible.

PROCESSES OF GROUNDWATER RECHARGE AND DISCHARGE

In general, groundwater recharge from the surface occurs after the soil moisture store is filled, when excess water may either run off as river flow or else drain down towards the water table as groundwater recharge; it is possible for both processes to occur as a result of recharge from stream beds. In some environments, as in chalk or limestone, the recharge will dominate the destination of the net rainfall. In some conditions, or in areas where rainfall intensity is high, infiltration through the soil surface may be a limiting factor, but in general the balance between recharge and runoff will depend on downward percolation rates from a full soil moisture store. Measurement or estimation of this phase of the hydrological cycle is not easy, but some examples are discussed later.

It is possible to assess the timing and quantity of recharge from a study of the fluctuation of the water table or piezometric surface, which can be measured in observation wells. This is particularly true where recharge is highly seasonal, because the rainfall or the potential transpiration is seasonal and therefore the net rainfall is seasonal. It should be noted that the rise in water level due to recharge is not the same as the depth of water recharged; the coefficient of storage is the ratio between increased water content and water depth of recharge, which depends on the aquifer material. In the case of a confined aquifer, where the water is held under a pressure which increases as the piezometric level rises, the storage coefficient is much smaller than in the case of a free water surface, typically 0.001 against 0.1.

The flow through an aquifer can be illustrated by a water table map, on which the contours of equal water table elevation are drawn. The direction of flow is perpendicular to the water table contours, from areas of higher to lower elevation. The rate of flow through the aquifer can be estimated by Darcy's law, which states that the flow is proportional to the gradient of the water surface and also the permeability of the aquifer material, which is expressed as cm day^{-1}, or as m^3 day^{-1} m^{-2} of aquifer, per unit gradient. This flow rate is not the same as the velocity of the water itself, because the flow pathway does not include the whole aquifer. The permeability may be estimated from the material of the aquifer, or tested in a laboratory, but can be measured more precisely by an *in situ* pumping test.

Groundwater flow will emerge at a lower level than the water table, either as river flow, spring flow or even through the coastline directly to the ocean, as sheet flow or through offshore wells. The groundwater discharge may be deduced by estimating the groundwater component of river flow, the outflow at springs or with more difficulty the outflow to the ocean.

SUMMARY OF TEHRAN STUDIES

The rate of groundwater recharge may be estimated or measured at different stages of the flow cycle: by estimating the recharge itself, the flow of groundwater through the aquifer, or by estimating the discharge from the aquifer to a river or spring. These three approaches are introduced by means of successive studies of the groundwater resources of Tehran.

Successive investigations of this major aquifer, described as the source of one of the largest developments in the world, have taken advantage of increasing data availability and advances in integrated modelling techniques. Initial information on aquifer recharge from rivers, on flow through the aquifer based on water table gradient and aquifer dimensions and permeability, and on seasonal water table fluctuations, was used to make preliminary estimates of groundwater availability.

As measurements of river inflows, well level fluctuations and aquifer properties were accumulated, together with estimated distributions of abstractions and aquifer recharge from external domestic supply, information became available for modelling. In the first stage an electrical analogue model was constructed to reproduce the average values of aquifer recharge, flow through the aquifer and abstraction. The analogues between flow through an aquifer and electrical current through a network of resistances, between aquifer permeability and electrical conductivity, and water level and voltage head, were used to construct a steady state model of the aquifer and to simulate wellfield abstraction at various locations and deduce the long-term effect on levels of different rates of pumping.

Once a further period of records had been collected, it was possible to assess a time series of recharge over a spatial grid from river flows and from water supplies; this information was included in a dynamic model of the aquifer, based on computer-based mathematical modelling. Water level data covering a seven-year period were summarised by a 64-node series, and recharge from streamflow and imported water, rainfall, *qanat* flows and abstraction were calculated for each node on a monthly basis. The mathematical model represents the response of the aquifer in terms of water levels and water flows to patterns of recharge and abstraction, given the aquifer characteristics of transmissivity and storage coefficient for each grid square. Transmissivity coefficients were first optimised over a period when cumulative level changes were small, and storage coefficients were optimised using the complete time-varying model. Once the model results were reasonable over the period of observation, the model could be used to assess the effect of wellfield abstraction over time. Thus increasing levels of information were used in a series of models of growing sophistication and precision.

Initial study of the Tehran aquifer

In 1957 when very little information was available in the form of measurements, it was necessary to estimate the annual quantity of water available from the local aquifer for supply to the city of Tehran (Sutcliffe & Rangeley, 1960b). The original water supply for Tehran was derived from underground resources by means of a number of subterranean galleries known as *qanats*. These (Fig. 6.1) consist of a tunnel sloping gently uphill from the point where the water supply is designed to emerge, upstream to a "mother well" below the water table in a source area. Vertical shafts are constructed at regular intervals to provide access for construction and repair, and these reveal the course of the *qanat* on air photography. The majority of the *qanats* emerged in the historic city core, where water was traditionally distributed in surface channels along the streets. At the time of the study the *qanats* had been supplemented by a number of drilled wells around the expanding city.

Further resources had been derived since 1933 from the River Karaj to the west by the construction of an aqueduct. The Karaj and another large river to the east of Tehran, the Jajerud, were being developed near their outfalls from the Alborz mountains by the construction of reservoirs, but the flows of these rivers were also used for irrigation on the plains to the south. It was therefore important to know what could be supplied from the local aquifer, which had already been developed to some extent by individual industrial and domestic water users.

At the same time as the hydrology of the aquifer was being investigated, the alluvial geology of the Tehran area was studied (Knill & Jones, 1968). The area of Greater Tehran lies between the volcanic deposits of the Alborz mountains to the north, the complex sequence of limestones and igneous rocks of the Sepayeh ridge to the east, and limits defined somewhat arbitrarily by the River Kan to the west and the city of Rey to the south. There is evidence that the ridges of the Sepayeh continue below the city, with a depression within the bedrock floor reaching a depth of about 700 m to the north of the city. The alluvium underlying the Tehran plain has been divided into four groups ranging from an older, folded group of probable Pliocene age, overlain by flat-lying Quaternary gravels, sands, silts and boulder beds. North of the city (Fig. 6.2) the

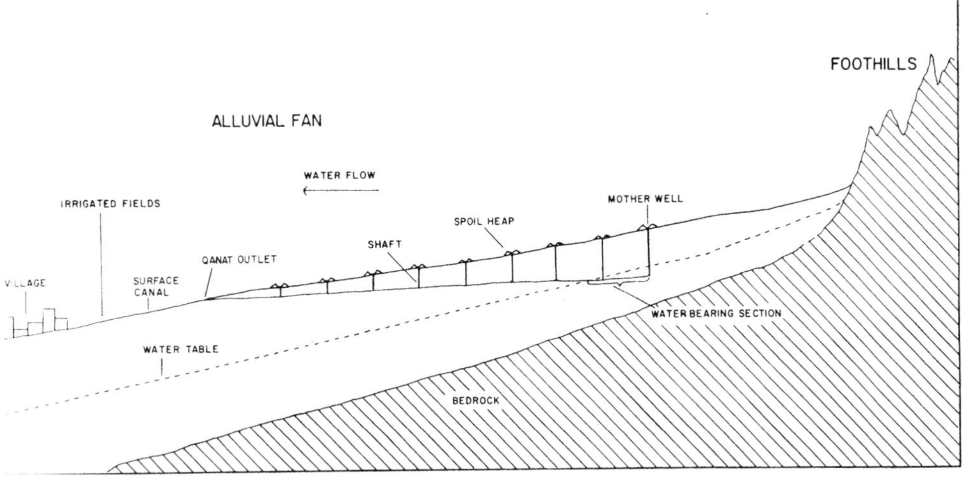

***Fig. 6.1** Diagram of a qanat (from Beaumont, 1971).*

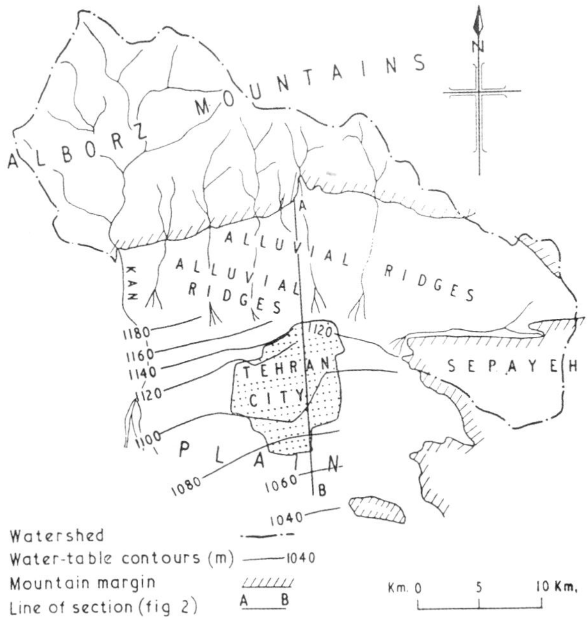

Fig. 6.2 *Tehran aquifer (from Sutcliffe & Rangeley, 1960b); cross-section A–B is shown in Fig. 6.3.*

plain is broken by a number of alluvial ridges, formed from the older beds contorted into anticlinal folds with steep or vertical bedding, which are aligned from east to west. The younger alluvial beds overlie the old beds unconformably and slope gently towards the south or form flat-lying deposits infilling depressions. Recent gravels occur within the river beds and their gravel fans.

The aquifer is extensive and unconfined within the alluvial beds underlying the Tehran plain. It derives water from a segment of the mountains to the north, draining towards the plain above the city. In the northern part of the plain a number of ridges of old folded alluvium with relatively poor aquifer characteristics, form a barrier to lateral movement of groundwater, which is restricted into passages between the ridges. Several watercourses cut through these ridges and spread out as wide gravel fans to the north and west of the city centre. The main water-bearing formations lie to the south of these ridges and slope gently towards the south and a number of salt lakes. The western boundary of the aquifer was taken as the River Kan (Fig. 6.2); to the west along the base of the mountains, this aquifer was used for local irrigation.

Estimates of recharge

A water balance study of the southern slopes of the Alborz range revealed a close relationship between precipitation and elevation (Fig. 3.6). The annual mean rises from 160 mm at an elevation of 1000 m to 630 mm at 2170 m. Precipitation occurs during the colder months (October–April) and the basin water loss is thus much smaller from the mountain area than from the plain. The main recharge of the aquifer derives from precipitation at high elevation, which runs off mainly as snowmelt through rivers and watercourses which debouch onto the plain. Several streams have cut through the alluvial ridges in narrow channels and all spread out into wide gravel fans and braided

Plate 6.1 *Alluvial fan of the River Farahzad, northwest of Tehran.*

channels once they reach the flat plain (Plate 6.1). Recharge clearly enters the aquifer through the gravel beds and fans. There was also significant artificial recharge in the area of the city from domestic and industrial sources.

The estimation of infiltration from surface sources was reasonably simple, because it was clear from reconnaissance that a large proportion of the mountain precipitation and runoff found its way into the aquifer. At the time the River Kan and the smaller streams were not monitored, but the precipitation on their local basins was estimated from the regional relation between precipitation and elevation. It was also possible to estimate the runoff from this precipitation by comparison of the water balance of several adjacent measured rivers, including the Karaj and Jajerud.

Some discharge measurements were made on the Kan during the study, and showed that a fairly high proportion of its flow passed the city in a defined channel; it was clear that much of the recharge from this river took place near the alluvial fan to the southwest of the city. The proportion of the Kan runoff recharging the aquifer to the east of the river above the city was estimated at 15% from measurements and observation. The smaller streams, on the other hand, were observed to fan out to the north of the city (Plate 6.1), and it was deduced that a large proportion of the runoff recharged the aquifer. Estimates were made (Sutcliffe & Rangeley, 1960b) of the recharge as a proportion of runoff; based on observation of the topography, these ranged from 40 to 80%. A nominal 10% of the precipitation, of about 200 mm, on the plain above the city was added to give a total estimated annual recharge of 67 $m^3 \times 10^6$.

In the unusual conditions of this semiarid area, this estimate provided a useful check on the subsequent estimates made by other approaches. The proportion of inflow reaching the aquifer must often be uncertain, though with rivers that disappear within the study area it was relatively simple. In other cases measurements of inflow and outflow could provide useful estimates.

Flow through the aquifer

At the time of the initial study (1957) no archive of well records or water-table map was available, so the first step was to compile these essential tools. With the assistance of local drilling contractors, some 225 existing wells were traced and visited. Initial depths to water table had been logged and were compared with the observations of the owners or mechanics of the wells, many of which were fitted with gauges. Levels at different dates were consistent over the area; a few records of depth over time were found, and one long-term record suggested that little change had occurred up to that time. This meant that the water-table map compiled from observations at different dates should be representative. Water samples were taken and analysed, while a number of estimates of seasonal fluctuations of the water table were obtained at the wells.

The detailed water-table contours showed that several individual groundwater flows enter the northern part of the main aquifer from different sources. Flows derive from the Kan River and from the fans near the Kan to the south of the alluvial ridges, and also pass through the gaps in the ridges to the south of the northern Shemran basin. The same areas provide the sources of most of the *qanats* supplying the city; these sources were located with the help of the city's chief *qanat* engineer and marked on air photographs and maps (e.g. Knill & Jones, 1968). South of the alluvial ridges the flows coalesce with a hydraulic gradient of about 1:120 (Fig. 6.3).

The information derived from water quality investigations can also provide useful hydrological insight by giving information on the processes at work. For example, samples of water obtained from over 100 boreholes throughout the Tehran plain were analysed; those obtained below the city, especially from relatively shallow boreholes, clearly illustrated the effect of downward drainage of wastewater by the presence of a mound of high total soluble salt values below the city. This artificial recharge within the city centre was also evident from a mound in the water table. Similarly, tongues of relatively fresh water through the Qolhak gap in the alluvial hills to the north of Tehran confirmed the routes by which groundwater flows through the aquifer.

Whereas recharge depends on surface water flows, which vary from year to year, groundwater flow through a large aquifer should be relatively steady, as annual

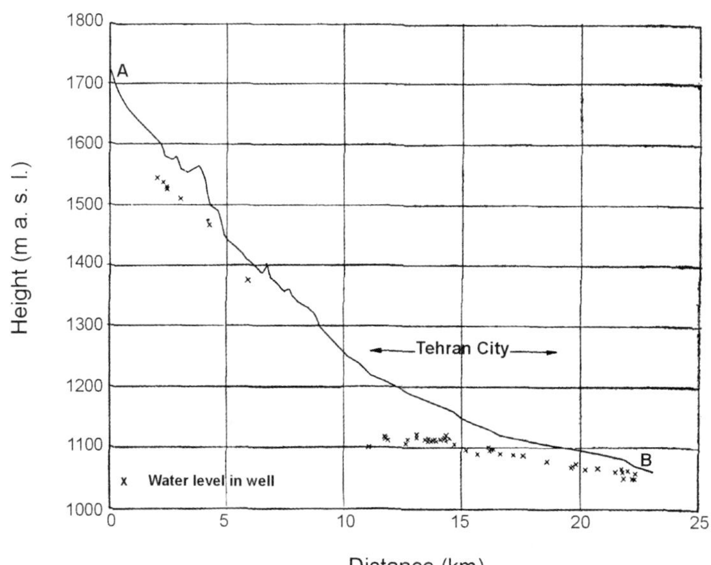

Fig. 6.3 *Cross-section of Tehran aquifer; location of transect A–B shown in Fig. 6.2 (from Sutcliffe & Rangeley, 1960b).*

fluctuations will be damped out within the aquifer. Thus estimates of flow through the aquifer, based on measurements of gradient and aquifer properties, should represent the long-term average resource. The information from a large number of wells showed that the hydraulic gradient in the main aquifer was uniform, which with the geological evidence suggested that the aquifer was fairly uniform horizontally.

This in turn suggested that the transmissibility of the aquifer, or the permeability multiplied by the depth, should also be fairly uniform and could be reasonably estimated from a small number of tests. The time for the study precluded the drilling of sets of wells for orthodox pumping tests, but during the visits to recently drilled well-sites, several were found where hand-dug shallow wells had been replaced by adjacent deep boreholes and modern pumps. The owners readily agreed to allow their wells to be used for testing.

Three pumping tests were carried out at sites of this type; the deep well was rested and then pumped for periods of up to 48 hours while the water table was observed in the nearby shallow wells. This was followed by an equal recovery period when the deep well was rested. In the absence of automatic recorders this entailed a shift system of observers using locally made electrical sounders (Plate 6.2), but the tests provided data which accorded with theory. The outputs of the pumps were measured either directly by V-notch or by volumetric methods. Drainage was a problem as the pumps were not normally used continuously; at one site where the output was passed into street drainage to avoid recycling, a check revealed that a small lake had been created to the delight of the local children.

The drawdown and recovery records at the observation wells were analysed using the graphical Theis curve and straight line methods (Wenzel, 1942; Brown, 1953). The plotted results (Fig. 6.4) fitted the straight line solution very well. In one case, in the south of the city, two observation wells were used at 21 and 108 m from the pumped well. In addition to the analysis of drawdown and recovery levels in each well, the final drawdown values were used for an equilibrium solution. The results for transmissibility

Plate 6.2 *Observing water levels during a pumping test northwest of Tehran.*

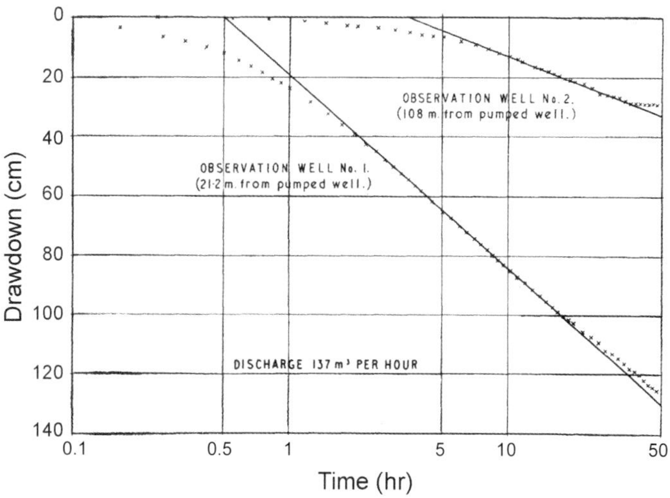

Fig. 6.4 *Pumping test in southern Tehran (from Sutcliffe & Rangeley, 1960b).*

were reasonably consistent except for those derived from the far observation well alone. A second test in the south of the city gave a similar result for transmissibility, while one in an isolated basin 10 km north of the city centre gave a lower value consistent with the steeper gradient in the area. Values of the coefficient of storage were also obtained from these tests. The typical value of transmissivity, in the unit current at that time, was 72000 Meinzer units (US gals/day/foot width/unit gradient) or 895 m^2 day^{-1}. The corresponding coefficient of storage was 0.1.

Because the pumped well did not extend to the base of the aquifer, the calculated transmissibility was adjusted to 1020 m^2 day^{-1} to allow for partial penetration of the aquifer, using depths given by geophysical survey. This transmissibility was multiplied by the hydraulic gradient and the effective width of the aquifer (20 km), to give an estimated annual groundwater flow of 62 $m^3 \times 10^6$. It was considered that this method of estimation was the most direct and involved the fewest assumptions. It is the most generally applicable method of estimating the available groundwater resources in a short period of investigation.

Water table fluctuations

Where recharge is highly seasonal, another method of estimation can be based on the observation of annual fluctuation of water levels within the aquifer. In the Tehran region, the precipitation occurs during the winter months, much of it in the form of snow. As the runoff is concentrated during the spring snowmelt, the recharge of the aquifer is highly seasonal and reflected in the fluctuations of the water table. Periods of recharge are followed by periods in which the outflow is equal to the rate of change of storage, which can be estimated from the product of the fall in water level, the coefficient of storage and the area of the aquifer. The rate of change of storage during the period of decline must be extended to cover the full year to estimate the annual recharge.

One or two records were available at the time of the study to confirm the seasonal pattern, but no network of measured wells had been established. However, questioning of mechanics at the wells revealed that where estimates were given, these were reasonably consistent and suggested that the seasonal range varied from about 4 m in the north of the main aquifer to about 1 m in the south of the city. As the recharge occurs during the first four months of the year and is followed by eight months of discharge, the seasonal range was increased by 50% to give the annual outflow. From the dimensions of the aquifer and a storage coefficient from pumping tests of about 0.1, the annual recharge was estimated at about 75 $m^3 \times 10^6$.

Comparison of estimates

The three estimates were based on different approaches: recharge from surface sources, flow through the aquifer, and water-table fluctuations. As far as possible, the three estimates were made independently, and yet gave annual figures of similar magnitude. Each of them could be criticised as depending on sparse evidence, but it was considered that a mean estimate of annual flow of about 70 $m^3 \times 10^6$ was reasonable.

At the same time it was recommended that, because the groundwater resource was important to the city, regular observations should be undertaken. It was suggested that the flows of the rivers and streams feeding the aquifer should be measured, that the River Kan should be measured at two locations, upstream and downstream of the main aquifer, and that a network of well levels should be monitored and systematic pumping tests carried out.

Development of analogue model

When the investigation was followed up in 1970–1973, sufficient records had been collected to provide estimates of the average behaviour of the Tehran aquifer, which was taken to cover a wider area than before. Estimates were made of the average rainfall over the Tehran region, the surface water inflow and the distribution of recharge, and of the water imported from the Karaj and Jajerud rivers and its distribution over water supply areas. Surface outflows to the south were also taken into account, and evaporation losses over the area were estimated. *Qanat* inflows and outflows were assessed, and were found to have increased significantly. An independent estimate of the transmissivity distribution was based on maps of recharge and water table, and a digital model was used to reconcile the transmissivity estimates, by providing successive transmissivity distributions which reduced the error in the water balance from that derived from pumping tests and geology.

These estimates were used to construct an electric analogue model of the aquifer (Cole, 1973). In this model the aquifer was represented by a two-dimensional network of points connected by resistances which reflected the transmissivity of the area, water levels were represented by voltage heads measured at these points, and groundwater flows were modelled by electrical currents through the network. The transmissivity distribution through the aquifer was derived from a number of pumping tests, extrapolated and adjusted to fit the hydrogeological evidence; the storage coefficients were derived from the tests and the aquifer lithology and were represented by capacitors. The levels at the boundaries of the aquifer were represented by voltages at corresponding points in the model, and the resistances ensured that the voltage network corresponded with the levels in the aquifer. The *qanat* system was represented by a number of "snake and ladder" circuits. Once the aquifer had been modelled in this way, it was possible to model the extraction of groundwater from a wellfield at any point by extracting a current at the equivalent point and observing the change in voltage which was equivalent to the drawdown. This steady state model of the aquifer, which was finally constructed on a 1 × 1 km grid, and provided with surface inflow and rainfall on the basis of an average year, enabled wellfield planning to be made without extensive trial drilling. The effect over time of wellfield abstraction could be tested on the analogue model. The effect of importing water or of diverting the River Kan could also be tested. Although it would be unusual to develop an analogue model nowadays, as computer-based models are relatively straightforward, analogue models have the advantage as an operational groundwater management tool in that they are comparatively simple to understand.

Development of dynamic model

About five years later again, the available measurements had accumulated to an extent that it was possible to derive a dynamic as opposed to a steady-state model of the aquifer. This time-varying model was designed to simulate the behaviour over a period of time rather than calculate an average condition. In order to make the adjustment of aquifer parameters simpler, computer-based mathematical modelling was used on this occasion. In order to monitor the behaviour of the wellfield, the aquifer was divided into 64 grid squares of 4 × 4 km, and the well levels recorded in 200 wells over a time span of 84 months were used to give monthly water levels for each grid square. The flows of the northern streams had been measured for seven years, and gaps in the record were filled by correlation with the flows of the larger rivers. Flow records of the streams at sites along the channels were estimated from gauged flows in order to deduce recharge from streamflow. It was possible to assess recharge on a spatial grid taking account of river flows in time and space. The import of surface water from the Karaj and the Jajerud had increased dramatically during the period of study, and the increasing water supply over the years from these imports and the resulting recharge to the aquifer were taken into account. Records of water supply were available from the supply network, and its distribution and use over the city could be deduced from sample billings which showed consumption. The residual water supply which was returned to the aquifer was deduced from a design study of surface sewerage systems, so that the inflow to the area, the recharge of the aquifer and the water level response of the aquifer could be compared with wellfield abstraction records and records of artificial recharge.

A number of channels draining from the lower end of the study area were measured and their outflows summarized. Recharge from rainfall, abstractions and discharges of *qanats*, and well abstractions, were calculated from periodic measurements or estimates and allocated over the nodes of the model and each month of the study period.

A dynamic mathematical model (Smith & Piper, 1978) was developed to represent the response of groundwater flows and well levels within the aquifer system to varying patterns of recharge and abstraction. Using the series of recharge and abstractions deduced above, the model reproduces the resulting changes of water level, using the parameters of the system: the transmissivity and storage coefficient of the aquifer over each grid square. The equations of groundwater flow and storage are represented by partial differential equations linking flows through the aquifer. The estimated flows depend on hydraulic gradients and aquifer transmissivity, while changes in storage are based on changes in head and storage coefficients, taking account of recharges and abstractions to and from the aquifer. In the model the continuous aquifer is represented by a rectangular grid of nodes and time is divided into discrete intervals. The partial differential equations are approximated by simultaneous equations expressing flow and storage changes between adjacent nodes and time intervals. These finite difference equations were solved iteratively to give a time series of predicted water levels for each node. These predictions, using the aquifer parameters found in the earlier study described above, diverged from the water levels observed at many of the nodes and a programme of optimisation was undertaken. Transmissivity parameters were first optimised, using a steady-state digital model over two periods when change of storage was minimal; a weighted mean of these parameter values was accepted. The differences between the initial and final transmissivity parameter values were expressed non-dimensionally, and the pattern of changes was mapped according to the grid of nodes (Fig. 6.5); the pattern of change was considered consistent with recent hydrogeological interpretation. Storage coefficients were optimised using the complete time-varying model. Scrutiny of the aquifer parameters was based on prior knowledge of the aquifer; limits were placed on the permissible values of the storage coefficient. The optimisation was based on defining an error function to measure the fit of the predicted to the observed water levels and then changing the parameter values to minimise the error function.

Once the behaviour of the aquifer over time had been satisfactorily reproduced in the model, the effect of various forms of exploitation of the wellfield could be studied and monitored in time as well as in the long-term average case.

Thus the accumulation of records of the different variables involved in the hydrological behaviour of the Tehran region enabled more sophisticated methods of analysis to be used. On the other hand it could be argued that the analysis was similar

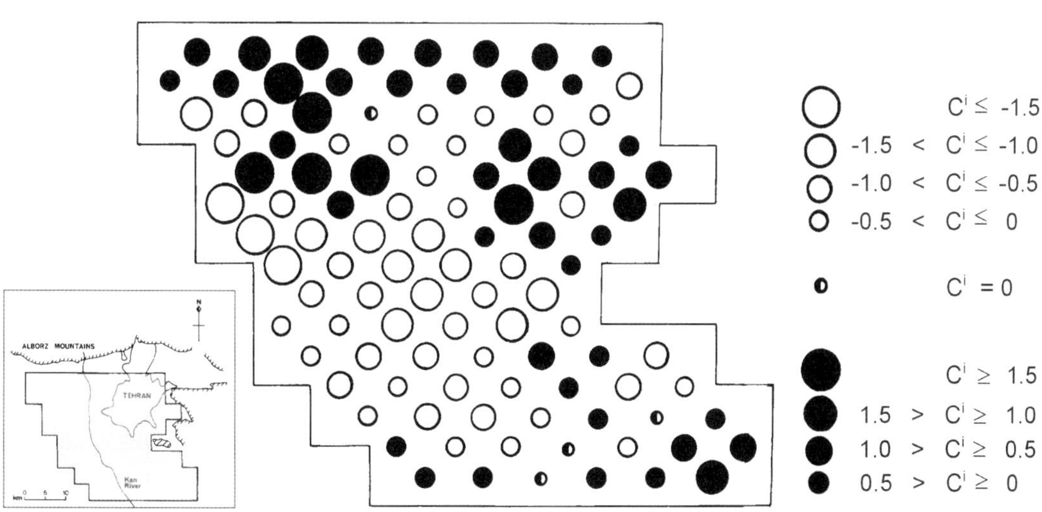

Fig. 6.5 *Non-dimensional adjustment of transmissivity values in the Tehran model (after Smith & Piper, 1978).*

to the original reconnaissance study of recharge, flow through the aquifer and water level fluctuations. The essential difference was that the various processes were studied in an integrated fashion, and the studies were based on actual observations of stream flows rather than estimates, records of well level fluctuations rather than verbal estimates, and a number of aquifer test results to enable the variation of the aquifer to be represented. Finite element models have now largely superseded finite difference models due to the ease with which grids of different sizes can be handled; nevertheless, the basic physical principles have not changed.

Example of joint exploitation of an aquifer

In 1957, at the same time as the initial study of the Tehran aquifer was being carried out, a complementary investigation was made of the water balance of the Varamin plain of about 1000 km^2, which is a continuation of the Tehran plain to the southeast (see Fig. 3.6). The interest of this example (see Fig. 6.6) is that it was in a less developed state than the Tehran area, and the water balance had not been disturbed from its historic state. It was fed by the River Jajerud, which drains about 1800 km^2 of the Alborz mountains and contributed about 400 $m^3 \times 10^6$ river flow annually at the apex of the alluvial fan. At the time preparations were being made to construct a reservoir at Latiyan on the Jajerud, to store water for the control of irrigation on the plain and to divert water to the city of Tehran.

Although the northern agricultural villages near the outfall of the river were supplied with irrigation water by a network of canals, the villages in the southern part of the plain were largely supplied by a series of *qanats* feeding a large number of small irrigation ditches. The whole area supported some 240 villages, which depended on wheat and barley, cotton and vegetables. At the time, only a few drilled wells had been established to provide water for industrial purposes, so the area was essentially in its natural state. As the average rainfall at Varamin was only 170 mm, the main contribution to the water balance was the river inflow. Estimates were made of the direct river contribution to the aquifer, and of canal seepage by means of successive flow measurements using locally-made Parshall flumes spanning reaches of canals, and also pond tests in which seepage was measured from short lengths of canal isolated by steel plates. Recharge from rainfall and irrigation was estimated.

However, the extraction of water from the aquifer was effected by over 200 *qanats*, and the balance could be estimated from their outflow. These *qanats* were typically about 15–20 m deep at the motherwells, and their channels had a gradient of 1:2000, compared with a ground slope of about 1:300; the length from motherwell to exit was about 2–5 km. Measurements of outflow were made at a sample of 100 *qanats* out of over 200 sites located from air photography. Velocities were measured by current meter (Plate 6.3), except for the smallest where floats were used; the measurements were made in late October, so the results were conservative, though seasonal variations were not thought to be extreme varying only with water table level at the motherwell. The total flow was estimated from these measurements at about 5 $m^3 s^{-1}$ or 160 $m^3 \times 10^6$ $year^{-1}$.

In general terms the hydrology of the Varamin plain was in equilibrium, with the river inflow being used successively for irrigation and the surplus draining into the aquifer by seepage from canals or irrigated areas and then being abstracted by *qanats* for further irrigation.

***Plate 6.3** Observation of qanat flow on the Varamin plain, southeast of Tehran.*

The area was studied in 1967, 10 years later, by Beaumont (1968). By that time the abstraction from wells had greatly increased; the total discharge was estimated at 145 $m^3 \times 10^6$ per year, and the water table was dropping from that recorded in 1957. He reproduced a map of the *qanats* from the 1957 air photography, which showed clearly the pattern of abstraction (Fig. 6.6). When he visited the area only two years later, in 1969, large numbers of *qanats* had been destroyed after heavy rains and had been replaced by well production (Beaumont, 1971).

The passage of water from the river channel of the Jajerud through the irrigation channels and *qanats* of the Varamin plain, and the successive use of water in irrigation and waste flow, until its emergence in the salt lakes of the desert fringe, were also illustrated in successive changes in water quality from low to very high salinity.

Variation of abstraction with time

This example of the Varamin plain from 1957–1969 illustrates the fact that the balance of a groundwater system may vary dramatically with time, especially during periods of technical change. In the case of the Tehran aquifer, it was possible to estimate the increase of abstraction over time, as the construction of the wells was relatively recent in 1957. Nearly all the wells had been drilled in the previous eight years, and the growth in abstraction could be deduced from the dates of construction. The diversions from the river had also been monitored, so that the increase in abstraction and import of water could be taken into account and the estimated flows of *qanats* included.

WATER RESOURCES OF BAHRAIN

In the case of the water resources of the island of Bahrain, inflow to the aquifer was supplied by historical groundwater recharge in the mountains of Saudi Arabia which depended on significantly higher rainfall in the distant past. The piezometric levels fall from northwest to southeast over the island, and these levels had been falling over the recent period before an investigation in 1967. One aim of the study was to determine the cause of the decline in levels; this involved a comparison with historical abstractions.

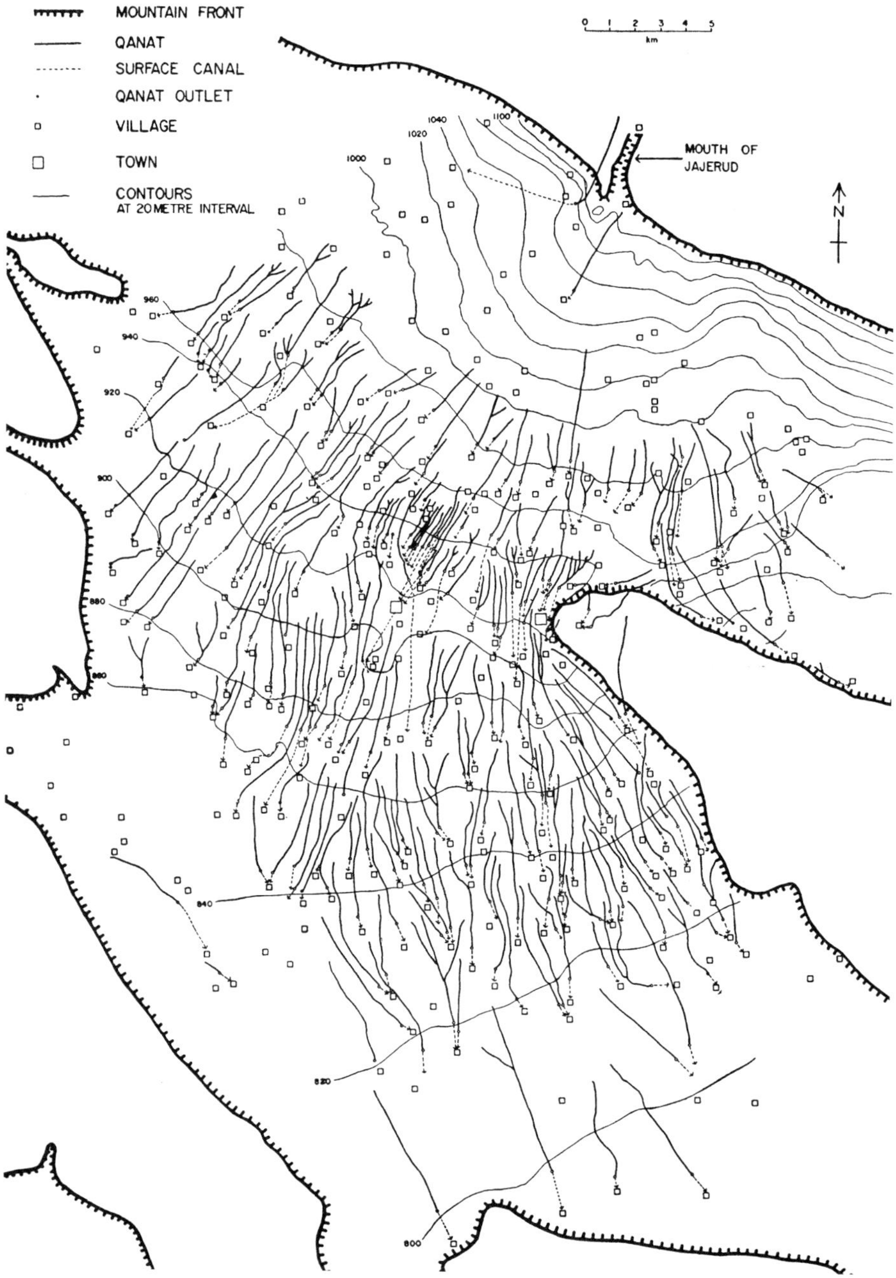

***Fig. 6.6** Varamin plain: contours and locations of qanats (from Beaumont, 1968).*

Fortunately, two previous studies could be drawn on to provide estimates for earlier dates. In 1923, before any well abstraction had begun, the only abstraction from the aquifer had been from natural springs, both on the island where date plantations were maintained from a few large springs whose yield had been measured, and from offshore springs whose output was measured subsequently. After exploratory drilling had revealed the potential of the aquifer, well construction had started in 1924 and had increased rapidly after 1947; a survey in 1953 had led to an inventory of wells and yields which provided historical evidence of abstraction based on the dates of construction. By 1967, the number of wells used for irrigation had increased to the extent that it was simpler to estimate the abstraction from the irrigated area, as described in Chapter 4. This series of snapshots of the abstraction could be compared with the decline in the water table.

A study of the water resources in 1967 focussed on the balance between the groundwater inflow, which derived from recharge in Saudi Arabia and flowed through the brown crystalline limestone Dammam Formation aquifer (Willis, 1967) under the sea, and the abstraction for domestic and urban use and for crop irrigation. It was necessary to deduce a history of water abstraction on the island for comparison with the declining groundwater levels. The early water supply had come from springs which provided drinking water and also supplied irrigation water to extensive date palm plantations in the northeast of the island. There were also submarine springs off the eastern coast which had been known for many years.

Several of the land springs had been surveyed and the yields of the main 15 springs measured in 1923. This information was compared with a wider survey of springs in 1953 to estimate a total annual yield in 1923 of 70 $m^3 \times 10^6$, with an addition of 10 $m^3 \times 10^6$ for the submarine springs. After 1924 a number of wells were drilled for water supply and irrigation, and a survey of some 300 wells had been carried out in 1953 to estimate the total abstraction. It was estimated that the annual well production was 60 $m^3 \times 10^6$, of which 70% was for irrigation. At the same time surveys of the land springs gave a total yield of 42 $m^3 \times 10^6$ and the submarine springs 10 $m^3 \times 10^6$ to give a total annual abstraction at that time of 112 $m^3 \times 10^6$.

By 1967, when another estimate was made (Sutcliffe, 1967), the number of drilled wells had reached almost 800, and the measurement of even a representative sample would have been a large task. The abstraction for urban and industrial water supply was measured, but the largest component was the abstraction from springs and wells for irrigation of date palm plantations. This component was estimated from the areas of irrigation and the transpiration from the date plantations, which was estimated to be similar to open water evaporation, as described in Chapter 4. Allowance was made for channel losses and leaching or drainage requirements. This estimate from the water balance, together with measured municipal and industrial use, led to the total annual abstraction of 122–147 $m^3 \times 10^6$.

The increased historic abstraction was further studied using the dates of construction and the positions of the wells drilled since 1946, and it was noted that the locations of abstraction had changed over the years. The effect of abstraction on Bahrain was reinforced by more recent abstraction in eastern Saudi Arabia. Comparisons suggested that the growth of abstraction was responsible for the recorded fall in well levels over the island.

The water quality was observed to deteriorate from northwest to southeast, in the same direction as the fall in water level. There was evidence that the water quality had also deteriorated over the years, and date plantations over parts of the island had been abandoned as a result. This has been attributed to leakage between the deeper aquifer and the sea through the overlying aquifer, leading to sea water intrusion as levels fell.

INDIRECT ESTIMATION OF GROUNDWATER RECHARGE

It is not easy to separate groundwater recharge from direct runoff as a residual of the rainfall:evaporation balance. One attempt to do this was made by Penman (1950) for the Stour basin above Stratford Mill in East Anglia, between Cambridge and Felixstowe, England. This basin is mainly sited on chalk with a variety of surface drifts, so that basin outflow is derived from groundwater rather than from direct runoff. Net rainfall could be assumed to recharge groundwater when the water content of the soil layer was at field capacity. He developed a simple model with a soil moisture store whose effective depth or root constant was related to vegetation rooting depth; for this purpose the basin was divided into areas of shallow-rooting grass and deep-rooting woodland, with root constants of 125 and 200 mm respectively, and riparian areas with unlimited supply. Net rainfall recharged each store and overflowed to groundwater when storage exceeded field capacity; when the soil moisture reduced to wilting point the evaporation was assumed to reduce to 10% of the potential rate. The estimated groundwater recharge over the period of record, 1933–1948, after river outflow had been deducted, was compared with the water level in a deep well in the chalk which had been observed over a number of years. It was found that the well fluctuations were closely related to the estimated storage (Fig. 6.7), with a lag of about two months, and a storage coefficient of about 0.01.

This simple model was tested on the experimental basin of the Ray at Grendon Underwood (Mandeville *et al.*, 1970), set up by the Institute of Hydrology, UK, to provide reliable records of a small basin. The study showed that it was possible to derive reasonable estimates of runoff from rainfall using this model. The Penman model also formed the basis of the estimate of mean soil moisture deficit in different parts of the British Isles (Chapter 5), which was used to estimate net rainfall in flood studies.

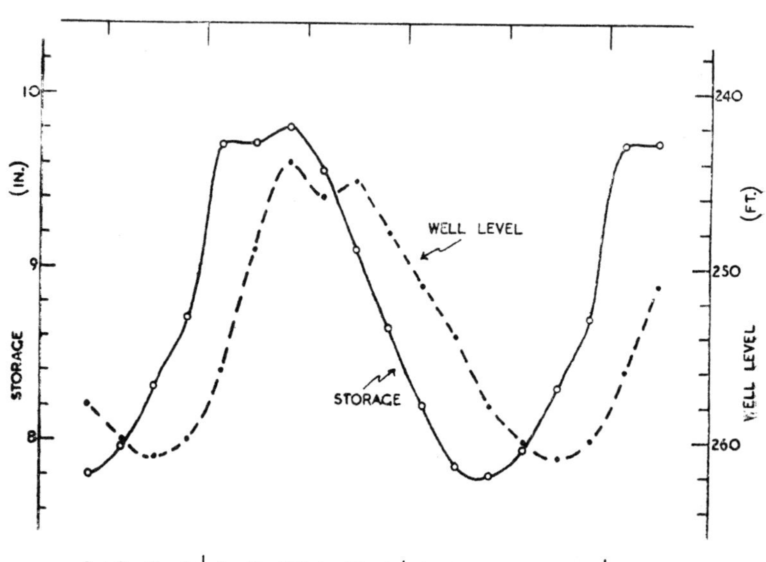

***Fig. 6.7** Stour basin: mean monthly values of storage and well level, 1933–1948 (from Penman, 1950). Note: 1 inch = 25.4 mm; 1 foot = 305 mm.*

Groundwater recharge in the Betwa basin

A similar model was used to study the water balance of the Betwa basin in central India (see Chapter 8), where it was hoped to estimate the groundwater recharge from the surplus; however, it was necessary to measure groundwater recharge directly as it was small by comparison with surface runoff.

In addition to the soil moisture network which was set up to monitor the water balance (see Chapter 5), two experimental sites were established in order to investigate the downward flux to groundwater directly. Experiments were set up on one site with deep soil (over 7 m) and one shallow soil site (2.4 m) (Hodnett & Bell, 1986). Duplicate sets of instrumentation were established (neutron probe access tubes, tensiometers at 0.2 m intervals to 2.4 m depth, and dip wells to measure water table depth). At each site, one plot was used to monitor soil moisture movement under natural conditions. The other plot was used to study hydraulic conductivity by the instantaneous profile method (Watson, 1966) with the plot covered by a black polythene sheet when the soil profile had been wetted to saturation. The hydraulic conductivity was calculated from the measured water flux and the potential gradient, and was related to matric potential and water content. It was shown that the conductivity decreases markedly with declining moisture content, and that downward flux is associated with cracks and fissures in the soil. Once the storage is filled by the rise of a perched water table, vertical percolation to deeper aquifers is very low. However, in shallow soils, especially under uncultivated grass and scrub where shrinkage fissures may reach the deep aquifer in weathered basalt, recharge is likely to be relatively unrestricted.

Thus it is possible to measure groundwater recharge directly, and such measurements can be important in establishing the processes involved in different environments. However, it is perhaps simpler to study recharge indirectly by measuring the aquifer response, especially in climates where the recharge is highly seasonal and therefore easier to observe. This technique was also used in the Betwa basin as well as the direct measurement based on soil moisture techniques.

Abu Dhabi and Oman

An extreme example where it was possible to deduce the water balance in a rapid reconnaissance, in February 1969, is given by a remote area (Plate 6.4) in the Al Hajar al Gharbi extension of Jebel Akhdar northeast of Buraimi/Al Ayn, on the border between Oman and Abu Dhabi. What could be termed "surface water villages" were concentrated along the watercourses where river flows could support permanently irrigated date palms and seasonally irrigated winter wheat. It was clear that the date palms relied on the baseflow or groundwater component of flow, while the wheat was irrigated from the seasonal rise. At lower elevations "groundwater villages" were located below the recharge sites where the groundwater flow through the alluvium was tapped by traditional tunnels (*aflaj*, sing. *falaj*, the Arabic equivalent of *qanats*) and wells.

A number of streams were observed to leave a range of volcanic rocks which are relatively impermeable, where soil formation and vegetation are almost absent. Most of the storm rainfall is likely to run off along the streams, where deposits of gravel and recent sediment provide storage which, with storage within the mountain range, appeared sufficient to maintain perennial flow where the streams leave the mountains

Plate 6.4 *Irrigated date palms within Jebel Akhdar, Oman.*

Plate 6.5 *An aqueduct carrying an irrigation channel over a stream, Oman.*

for the alluvial plain. Near this point there was usually a village relying on surface irrigation channels leading off the stream above a small stone and mortar embankment, which apparently had to be renewed after floods. These channels were well constructed and included aqueducts (Plate 6.5) to cross tributaries. Each village had a dense plantation of date palms, with winter cereals growing in fields downstream which were

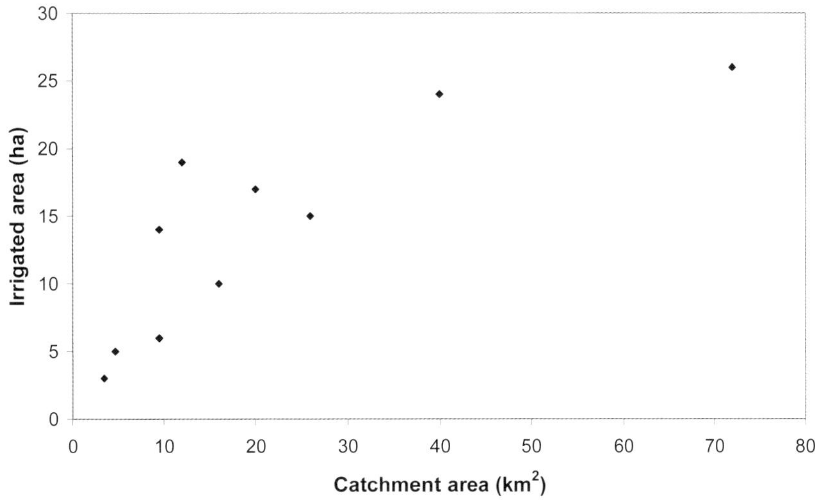

Fig. 6.8 *Oman: comparison of basin areas and irrigated areas.*

irrigated from extra flow available seasonally. Below most of these surface water villages were terraced fields which could be cultivated in years of heavy rains. There is a reasonable relation between the perennial irrigated areas of these surface water villages and the catchment areas upstream (Fig. 6.8).

Once the streams leave the main range they tend to enter deep gorges, up to 30 m deep, incised within the gravel plain which slopes gently to the west. Series of sedimentary deposits were exposed in the sides of these gorges, and the flow which passes the first set of villages is likely to percolate rapidly into these sediments. Once the streams reach the gravel plain the beds become difficult to distinguish in the mass of cobble outwash. To the west of this plain the channels draining the main range pass through an outcrop of limestone ridges and it is evident that floods pass through these channels. A small number of larger villages, including Buraimi and Al Ayn, are located within these gaps. These western or "groundwater" villages tap the subsurface flow by means of tunnels or *aflaj* which have their sources in or near the *wadi* bed. The flow is distributed in lined channels to date plantations. In the case of Al Ayn the total yield of seven *aflaj* was measured and found to be about 320 $l\ s^{-1}$ or 10 $m^3 \times 10^6\ year^{-1}$, corresponding to a depth of 3.2 m over the date plantations of 3.15 km^2.

In this arid area it was evident that most of the runoff had its source in the main range and that most groundwater recharge would occur during floods at sites where the watercourses left the mountains for the alluvial plains. At the time there were no measurements of rainfall or runoff, but it was possible to establish that there was also a reasonable relation between the total area irrigated in the eastern and western villages depending on specific streams, and the area of the main range contributing to the total runoff. The irrigated areas could be assumed to use about 3–4 m of water, which was equivalent in terms of the linear relation to 40–55 mm of runoff from the mountain areas. The rainfall in the mountains was likely to exceed the rainfall at Sharjah, which averaged about 100 mm, but it was clear from this brief reconnaissance that a significant proportion of the likely runoff from the mountains was already being used. In other words an efficient system of water use had developed over the centuries, with villages using surface and sub-surface water in two stages: at the margin of the main range and subsequently in natural gaps in the ridges downstream. The combination of perennial and annual crops no doubt make it easier to adjust to variations in water supply. Recommendations were made for the measurement of rainfall and runoff at selected locations, but it was clear from a visit to the area, even before measurements could be made, that there were unlikely to be large unused resources in the area.

Subsequent detailed investigations were later undertaken in another sector of the Jebel Akhdar, including rainfall and runoff measurements in *wadis* and *aflaj*. A mathematical model of *falaj* behaviour was developed (Parks & Smith, 1983) which was used to compare rainfall data with modelled outflow from a *falaj* whose channel and aquifer dimensions were known. This suggested that recharge occurred along the whole *wadi* channel following rainfall in the highlands and a flood wave down the *wadi*. The flow down the *wadi* quickly becomes exponential, with a recession constant related to aquifer dimensions and characteristics; thus flow modelling can provide similar information on aquifer characteristics to a pumping test. The important difference between these *aflaj* and the *qanats* studied near Tehran and the Varamin plain was that these *aflaj* drew on relatively narrow bands of groundwater flow. The *qanats* of the Varamin plain were able to tap a much wider aquifer which was supplied from a major river. Nevertheless the principle of aquifer abstraction is similar.

CONCLUSION

The processes involved in groundwater recharge are as amenable to water balance analysis as other components of the hydrological cycle. Moreover they may be approached in several ways: as the outcome of a recharge from soil moisture storage or from stream inflow; from the estimation of flow through an aquifer; or from evidence of seasonal water level fluctuations in an aquifer. The analysis may be based on a simple balance estimate, or on modelling of a stationary process, using average values, or on dynamic modelling of a variable process.

7 RIVER FLOW

River flow is the only component of the hydrological cycle which can be measured rather than sampled. Therefore the river flow station is particularly important in the hydrological investigation of a basin or an area. It is essential to ensure that flow records are as precise and consistent as possible. In this chapter some of the techniques of quality control are described and examples given to illustrate likely problems.

USE OF RIVER FLOW RECORDS

The records from flow stations provide the variation over time of the water available at the outfall of a basin. These flows may be used to derive average flows and describe the variability at different periods of the year. Low flows may be compared directly with water demands for run-of-river schemes, while cumulative totals give the volume available for storage where natural flows are inadequate to supply demand throughout the year. The measured flows may be extended to cover a longer period by comparison with adjacent flow records or with rainfall records. Although the accuracy of records is desirable, it is for this purpose that the consistency of records is equally important.

Apart from comparing supply and demand, river flows provide important information about the regime of a basin. Comparison with rainfall provides information on the water balance; a diagram of the seasonal flow gives a simple picture of the regime of the river basin; a typical recession curve, which can be derived from the seasonal flow diagram, gives information about the groundwater regime of the basin; the shape of the flood hydrograph, together with information on the storm magnitude and time distribution of the corresponding rainfall, allows the flood potential of the basin to be estimated.

BASIS OF RIVER FLOW MEASUREMENT

River flows are measured in terms of water level at a site where the level can be directly related to the rate of flow. This site may be at a structure which can give a precise relation between upstream level and flow over a weir; the structure may be designed for this purpose, or may be a structure designed to control the river level which can be adapted to measure flow. The upstream level is converted to flow using a theoretical equation for the structure or a relation derived from measurements. This approach is common in Britain where the river channels are usually small enough for this approach to be economic.

In other cases the river levels are measured at a site on a natural river channel where the relation between level and flow is likely to be regular and stable, for instance above a natural control such as a rock outcrop. In this case the relation between river level and flow is built up by a series of discharge measurements over the range of river level variations. Each discharge measurement consists of a number of velocity observations across the channel, using a current meter at a fixed fraction of the depth, or more rarely sampling the depth profile. The product of velocity and channel depth is integrated to give a discharge linked to the water level at the time of observation. The relation between level and flow is built up into a rating curve from the series of discharge measurements, either by plotting on natural or logarithmic graph paper, or by statistical correlation between level and flow. Account is taken of changes in this relation because of channel changes, so that each rating is linked to a specific period and range of level.

Types of river flow stations

A number of rivers in Britain are small enough to be measured using calibrated structures and automatic recorders, reducing the amount of fieldwork needed to maintain a gauging station; however, in many developing countries the rivers are larger and it is necessary to carry out discharge measurements and to calibrate the station using a rating curve derived from these measurements.

In semiarid countries the periods of river flow are so intermittent that it is difficult to carry out sufficient gaugings to build up a rating curve; in these conditions there are advantages in using a structure, if one can be designed to cope with the sediment and bed loads which are associated with the extreme variability of flow. For example, in the case of recharge to the Abu Dhabi aquifer from the Jebel Akhdar in Oman, discussed in Chapter 6, simple weirs were proposed to measure the flows of the *wadis* because of the remoteness of the potential measurement sites.

Because of the need for long-term flow series to sample the variations in climate which persist over relatively long periods, especially in more arid regions like much of Africa, consistency of flow records over the years is extremely important. Although methods of measurement, calculation and analysis are more complex for flow records than for some other variables like rainfall, homogeneous records are equally important and perhaps more difficult to achieve. The maintenance of complete long-term records at the same site and with consistent methods of discharge measurement, derivation of rating curves, and subsequent analytical methods, should be given a high priority.

Areal sampling

The question of areal sampling was discussed in Chapter 2, but it is useful to recapitulate some points here. In most countries it is not possible to measure the flows of all rivers, but it is necessary to design a network which could be used to estimate flows at ungauged sites, by sampling a wide range of climate, topography, geology and basin size. It should be possible to sample areal variability of runoff as effectively as time variations. Although annual runoff is largely related to basin area and the depth and seasonal variation of rainfall, the distribution of runoff through the year is influenced by the size, the soils and geology of the basin as well as the seasonal distribution of rainfall; all these factors should be sampled by a gauging network.

The number of gauges required will depend on the variability of these factors over a region. One way of establishing the adequacy of a network is by developing statistical relations between, for example, mean annual runoff or the mean annual maximum flood, and basin characteristics like size, rainfall and other factors. The more characteristics needed to give a reasonable estimate of flow factors, the more extensive a network is desirable. In practice, the size of a hydrometric network is likely to be affected by the importance of water resources to the national economy, and thus by the number of stations which can be maintained by the allocated budget. Within these restraints, a combination of long-term stations to sample time variations in combination with rainfall stations, and a well distributed network to sample areal variations, will give the optimal basis for water resources assessment for a given budget.

Quality of individual records

Once the network has been designed to sample flow variations in time and space, it is important to maintain the reliability and perhaps even more the consistency of records. As far as possible the whole series of flow records should be of a uniform standard of reliability. Where this is not possible, for instance because of changes of technique or difficulties of measurements in certain periods, it is important that this should be recorded so that this is taken into account when analysing records.

The measurement of river levels is likely to vary according to circumstances, but its frequency should be related to basin size and water level variability. The conversion of water level series to river flow series largely determines the reliability of flow records, and this conversion depends on the rating curve used at the individual site. The precision of flows depends largely on the stability and reliability of individual rating curves, and these depend not only on the methods used for discharge measurements and for deriving curves from them, but also on the physical characteristics of the site itself.

Methods of discharge measurement are reasonably standard and well described (e.g. ISO, 1997b). Velocities are measured at a number of verticals across the river (Plate 7.1) at standard fractions of the depth, using current meters which should be recalibrated regularly. In general measurements are made at 0.6 depth, 60% of the depth below the surface where the mean velocity is assumed to occur, or at 0.2 and 0.8 depth, where the average of the two observations is assumed to correspond to the mean velocity. In Nile basin practice, on the other hand, measurements have been made at half depth and the velocities multiplied by 0.96. Where possible, flows are measured using structures, either designed for the purpose or adapted from other structures. These may be calibrated from theory or from measurements. In general measuring weirs and flumes are limited to smaller rivers, but releases from reservoirs are usually monitored in this way.

At sites where gauging by current meters is not practicable, especially at high flows, velocities may be deduced from surface floats or by hydraulic formulae using cross-section surveys and gradient measurements. However, this is not as simple as it sounds; during flood events in New Zealand, it was extremely difficult to reach gauging sites during peak conditions, let alone attempting to record the flood profile with survey pegs. However, the debris left by the flood provided a consistent set of levels which could be used to estimate the flood discharge using Manning coefficients derived from moderate flows.

DERIVATION OF RATING CURVES

Rating curves, linking water levels to river flow at a given site, are derived from discharge measurements and thereafter used to estimate the river flow at that site from measured levels. The rating curve is thus the key to precise river flow estimation, but

Plate 7.1 *Gauging station, including staff gauges and a cableway, on the River Indravati at Jagdalpur, central India.*

the derivation of rating curves from discharge measurements has not received the attention it deserves. Several questions arise. Should the fitting of a curve be carried out by a hydrologist familiar with the site, or should it be based on an automatic statistical procedure? Should the rating curve be based on the gaugings of a single year, which is common practice in many countries, or on those of a number of years? The answer to the second question may depend on the purpose of the study or the characteristics of the site. For example, most of the records used in the investigations leading to the *Flood Studies Report* (NERC, 1975) were based on high flow measurements over a number of years, as these measurements were rare and the channel control was assumed to be more stable than at low flows. In a study of the outflow from Lake Kyoga in East Africa, a similar approach was taken, with the use of several years of gaugings as the range of lake-fed water levels was insufficient in most single years to determine the rating curve.

Traditional methods of fitting curves to points representing individual gaugings have been graphical, with curves drawn on either linear or logarithmic graph paper taking account of the geometry of the site. More recently, with the advent of computer-based techniques (Institute of Hydrology, 1993), equations of the form $Q = a(h - h_0)^n$ have been fitted over the whole or part of the range of levels. The level h_0 corresponding to zero flow and the exponent n are chosen to minimize the scatter, while bearing in mind that h_0 and n are both related to site geometry; h_0 may correspond with the level of rock control while n should generally be between 1.5 and 2.5. With modern programs, curves may be displayed on screen as linear or semi-logarithmic curves, with successive gaugings grouped and labelled. This enables the hydrologist to consider first whether all records are homogeneous by producing one rating curve to cover the whole period; next all the evidence is available to test whether the rating changes with time and to experiment with different groups of gaugings corresponding with time-varying rating curves; each curve may be divided into several sections where a single equation is not appropriate for the geometry of the whole channel cross-section; finally the need for separate treatment of rising and falling river levels may be examined.

Examples of rating curves from the Nile basin

During an assessment of hydrological records in several countries in eastern Africa, attention focused on the quality of discharge measurements and particularly on rating curves. Gauging techniques were studied and the ways in which gaugings were combined into rating curves. The resulting flow records were tested by double-mass analysis with other stations, and by water balance comparisons with basin net rainfall estimates. Thus a number of stations in the Nile basin were examined in the course of the study; an outline map of the basin (Fig. 7.1) illustrates the relative positions of these stations.

These examples from the Nile conveniently illustrate the various types of rating curve, as most variants are available. Some ratings are reasonably precise and stable, others are precise over periods of a year or so but vary over time; yet others have looped relations, where the relation on a rising curve differs from that on a falling curve. In other cases it is preferable to derive ratings from a number of years, as the range is inadequate in a single year.

The outflow from Lake Victoria was historically controlled by the geometry of the Ripon Falls, where the relation between upstream lake level and lake outflow depended on rock outcrops which formed sills and islands which were stable during historic times. In order to establish a rating curve for the lake outflow, discharge measurements were made from 1923 at Namasagali, a site some 80 km downstream of the lake, and were related to water levels at Jinja, near the outfall from the lake. This relation was well defined by the gaugings. After 1951 the construction of the Owen Falls dam, some 3 km downstream of the Ripon Falls, was begun with the construction of a coffer dam. Thereafter the flow was controlled by agreement between Uganda and Egypt, to accord with the natural flow based on the lake level and the rating curve, at least over periods of about 10 days to a month. This "agreed curve" is illustrated in Fig. 7.2, together with gaugings made between 1923 and 1950. After 1961–1964, when the lake level rose dramatically and remained above the limits of previous gaugings, the curve was extended from the earlier curve by hydraulic modelling. The gauging site was moved some 20 km upstream from Namasagali to Mbulamuti, to avoid the disturbance from the rise of Lake Kyoga downstream. The gaugings were supported by measurements of turbine flows and sluice releases past the dam. It will be seen later that the rating curve is confirmed by comparison with downstream flows.

Another station with a stable site supported with discharge measurements is the Blue Nile at el Deim, about 85 km upstream of Roseires near the border between Sudan and Ethiopia. This station was established in 1962 during the construction of the Roseires dam, and discharge measurements have been carried out regularly since that date. Figure 7.3 illustrates the rating curve and the gaugings on which it has been based; it is clear that the rating is precise and has not changed over the period of some 30 years.

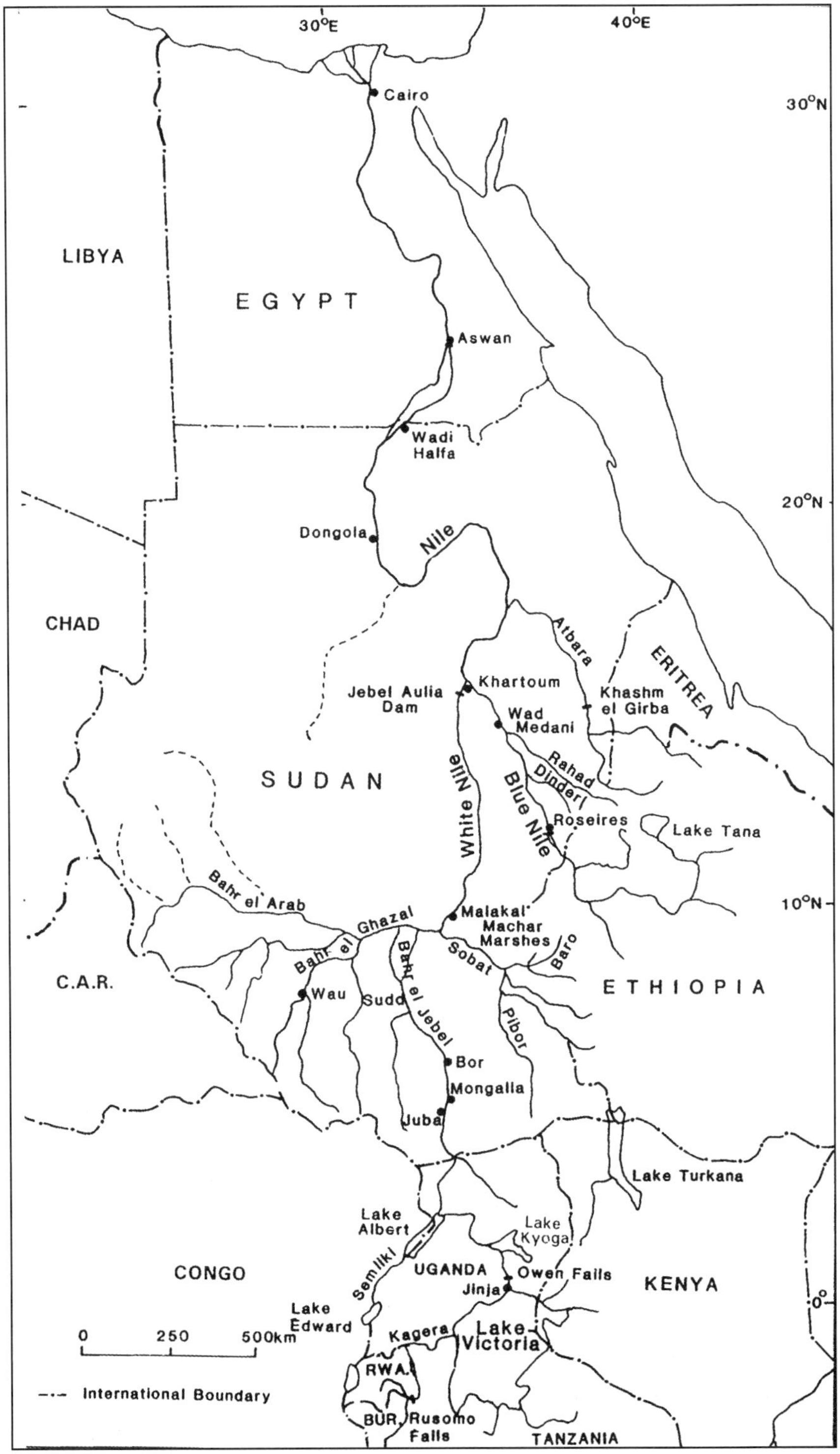

Fig. 7.1 *Map of the Nile basin showing key stations.*

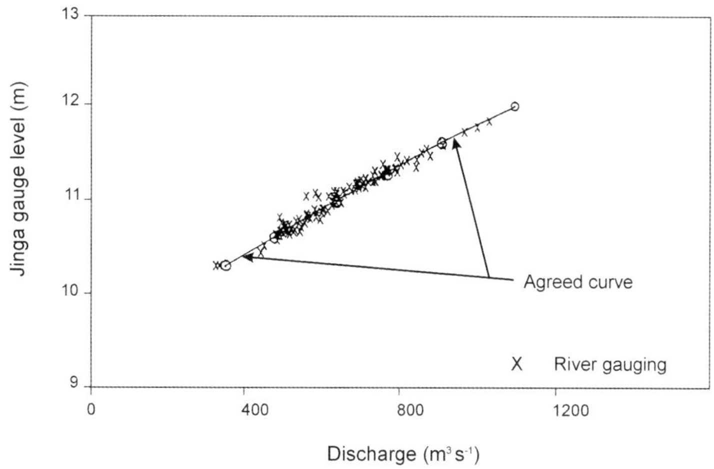

Fig. 7.2 Outflow from Lake Victoria: "agreed curve" and gaugings at Namasagali (after Sutcliffe & Parks, 1999).

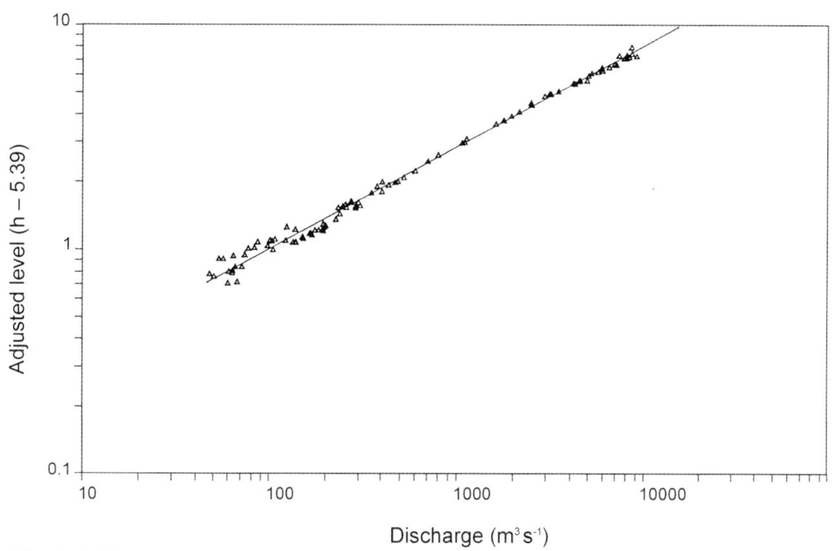

Fig. 7.3 *Blue Nile at El Deim: rating curve and gaugings, 1962–1981.*

Although the outflows from Lake Victoria can be derived from lake levels measured since 1896, and a combination of the rating curve and other measurements, the outflow from Lake Kyoga downstream has been less precisely monitored. Although levels have been measured at Masindi Port below Lake Kyoga at a site which was believed to have a stable rating, subsequent measurements revealed that the rating changed at intervals. During an investigation of flows between Lake Victoria and Lake Albert (Sutcliffe, 1996), rating curves were derived for Kamdini, some 100 km below Lake Kyoga. Levels were available from 1940 to 1980, while gaugings had been carried out at Kamdini itself from 1940 to 1959. However, gaugings were also available downstream at Fajao and Paraa, between Kamdini and Lake Albert, from 1940 to 1979, and there are no significant tributaries between the two sites.

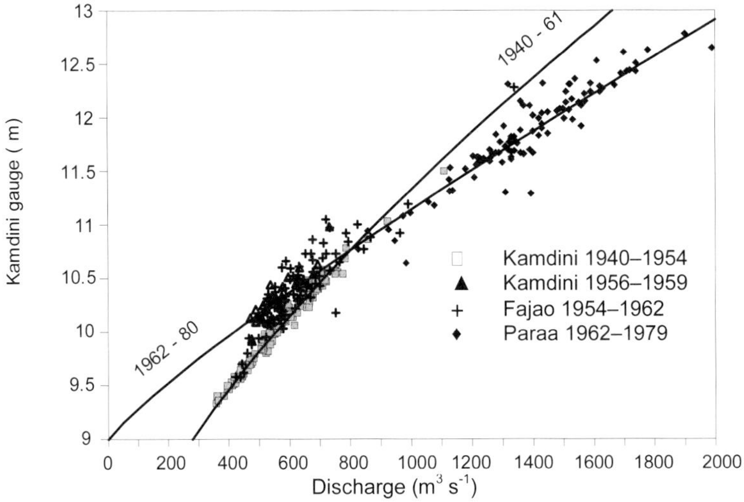

Fig. 7.4 *Kyoga Nile outflow at Kamdini: gaugings at Kamdini, Fajao and Paraa, 1940–1980 (from Sutcliffe & Parks, 1999).*

Because the levels at Kamdini are dependent largely on outflows from Lake Victoria, there is insufficient range of level or flow in a single year to define the rating curve. However, it was possible to derive a family of curves from gaugings at Kamdini itself and at Fajao and Paraa, related to levels measured at Kamdini. These curves, which appeared to be reasonably stable over one or more years, differed only in the level corresponding to zero flow, which was derived from all the gaugings in the period. After 1961, when Lake Victoria rose and the outflow more than doubled, there was also a change (Fig. 7.4) in the slope of the curve. Thus a series of curves could be derived to convert the levels to flows in this reach.

The next long-term station downstream of Kamdini was established in 1905 on the Bahr el Jebel at Mongalla, where the inflow to the Sudd is measured in a single channel. Although gaugings were infrequent between 1905 and 1921, river levels were measured regularly and were converted to flows using a general rating curve based on all measurements during this period. After 1922 measurements were frequent, with an average annual total of 260 between 1922 and 1931. Gaugings continued reasonably frequently until 1978, but were discontinued after 1984. Successive rating curves over the years show that the level corresponding to a given flow rose steadily (Fig. 7.5) from the beginning of the century until 1964, when it fell abruptly. This pattern was confirmed by the rise in the bed level derived from gauging records. The detailed change after 1963–1964 is illustrated in Fig. 7.6, where a fall in effective level of about one metre is shown. These changes were doubtless related to the fall from high Lake Victoria levels at the end of the last century, and a reversal after the sudden lake level rise in 1961–1964.

A different problem is illustrated by rating curves at Khartoum, where there is a marked difference (Fig. 7.7) between the gaugings during periods of rising levels and those of falling levels. There is a clear loop between the ratings for the two periods. Similar hysteresis effects are shown at other sites near junctions, like those of the Sobat and White Nile. At Khartoum the picture is complicated by an apparent change in

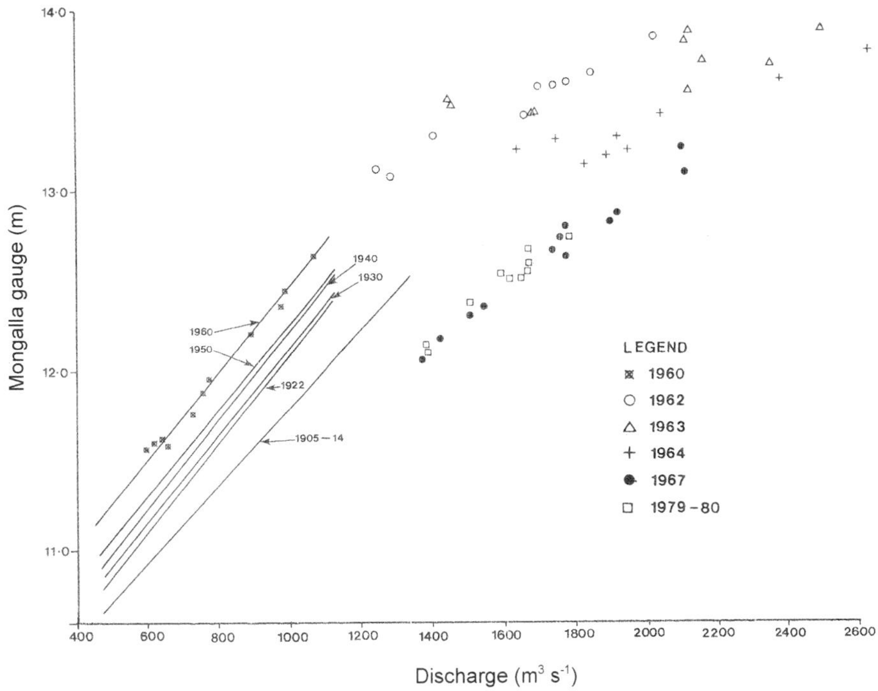

Fig. 7.5 *Bahr el Jebel at Mongalla: gauge discharge relations, 1906–1980 (from Sutcliffe & Parks, 1982).*

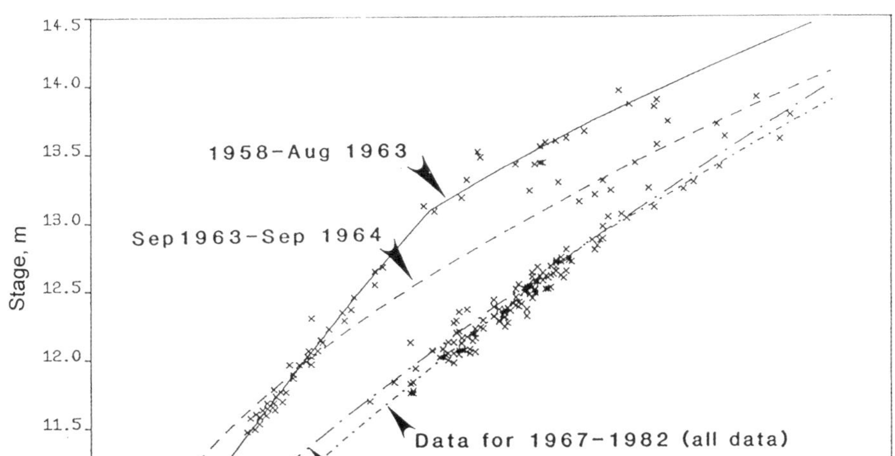

Fig 7.6 *Bahr el Jebel at Mongalla: detailed change after 1963/64 (after Sutcliffe & Parks, 1999).*

ratings at high flows, where a rise of about 0.5 m between 1902 and 1982 (Fig. 7.8) has been deduced from an analysis confined to the highest gauging carried out in each year.

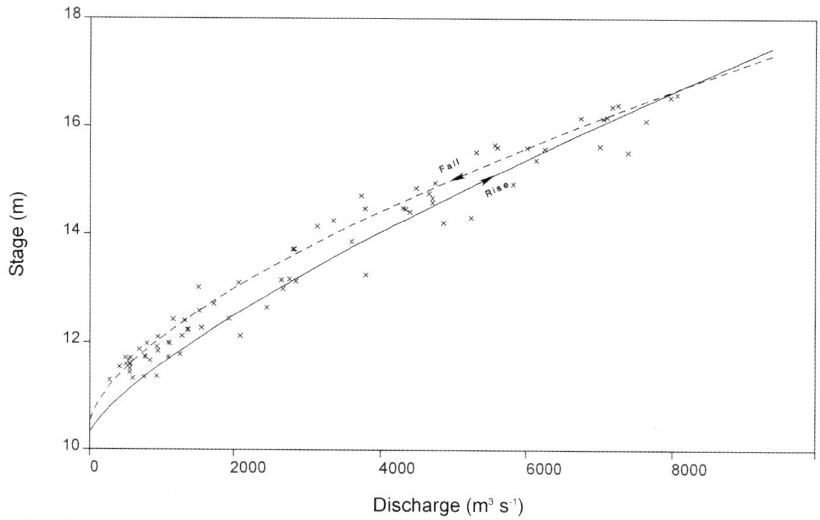

Fig. 7.7 *Blue Nile at Khartoum: rating curves on rising and falling levels generated using data for 1959, 1969 and 1980.*

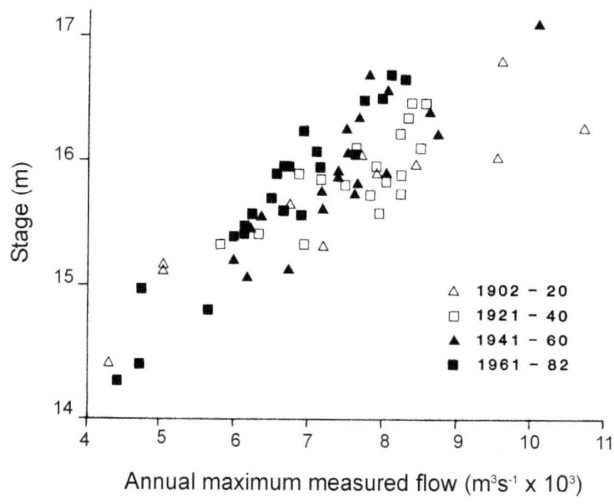

Fig. 7.8 *Blue Nile at Khartoum: ratings at high flows (from Sutcliffe et al., 1989).*

These examples from a single river basin illustrate a number of different types of relation between level and flow. The differences depend on the stability of the river profile and also on upstream and downstream conditions. It is important that the rating curves at specific sites should be reviewed over the years. This has been made easier by modern techniques of analysis which reduce the time necessary for routine computation and thus make it easier to compare gaugings and ratings over the whole period of record.

INDIRECT METHODS OF APPRAISAL

There are a number of indirect ways in which river flow records can be assessed for precision and consistency. These include double-mass curves, which provide a graphical comparison of different records, and statistical methods of comparison.

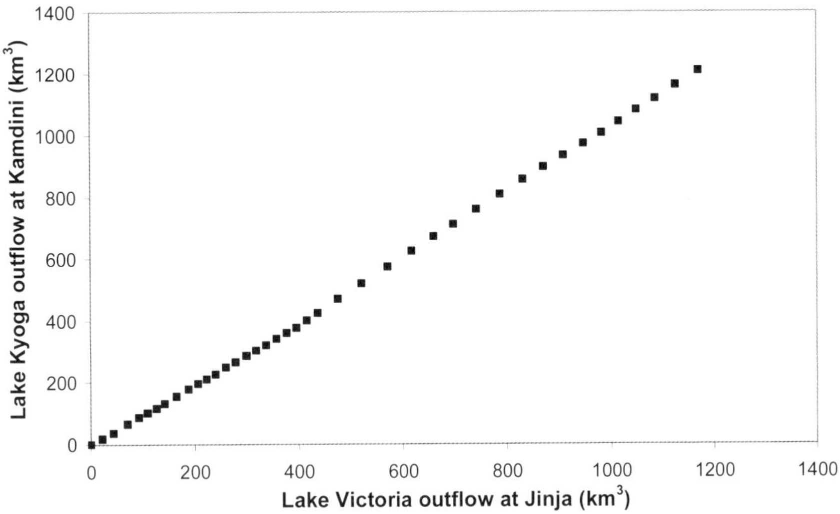

Fig. 7.9 *Cumulative flow comparison: Lake Victoria outflow at Jinja and Lake Kyoga outflow at Kamdini, 1940–1980.*

In general double-mass curves are the most popular form of comparison. The cumulative flows of different records are compared, or alternatively cumulative flows are plotted against time. An example comparing the outflows from Lake Victoria at Jinja with Lake Kyoga outflows at Kamdini (Fig. 7.9) suggests that both sites provide reasonable and consistent records. However, comparisons between stations on different tributaries may reflect different regimes rather than inconsistent records. For example, comparisons between the flows of the Bahr el Jebel at Mongalla and the Blue Nile at Khartoum show a distinct change after 1961–1964, but this is related to the rise in Lake Victoria and the doubling of outflows down the White Nile which was not reflected in the flows of the Blue Nile.

Double-mass curves can also be used to compare rainfall and river flow records, and can be useful in a number of cases to confirm the validity of records. However, where there is a discontinuity of rainfall and thus runoff over a period of years, there may be an apparent change in the double-mass curve. This is illustrated by the records of the Kagera which flows into Lake Victoria as its largest tributary. The Kagera basin was subject to an increase in rainfall after 1961, which also affected the rest of the lake basin; however, the runoff increased more than the rainfall in relative terms, and therefore the cumulative comparison of the two factors showed a distinct change of gradient (Fig. 7.10). This is of course due to the fact that the difference between rainfall and runoff is due to evaporation, so that runoff, as the residual, is extremely sensitive to changes in rainfall. If the rainfall over a basin is 1600 mm and the evaporation 1500 mm, the runoff will be 100 mm; if the rainfall increases by 10% without any change in evaporation, the runoff will increase to 260 mm, and the double-mass curve will show an apparent discontinuity.

One example where a comparison between rainfall and runoff provided the date of a discontinuity in level measurement, which was reflected in the flows of the Tongariro at Turangi in New Zealand, has already been described in Chapter 2. An example where such a comparison provided an assurance that flow records were valid was met

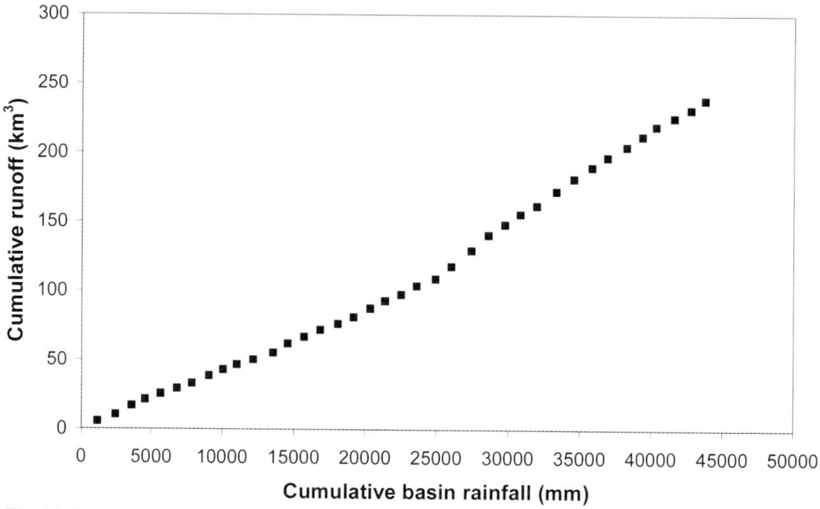

Fig. 7.10 *Kagera basin: cumulative rainfall and runoff, 1940–1977.*

in a similar climate on the fringe of the Andes in Argentinian Patagonia. Here the inflows to a hydroelectric station after the construction of the dam were lower than in the previous period of recorded flows. However, double-mass comparisons of flow records with independent gauging sites showed that the two sets of records were consistent (Fig. 7.11), but also comparisons with long-term rainfall records confirmed that there had been a significant decrease in the regional water balance. This comparison was made easier by the fact that the construction of the reservoir would not in itself have affected the runoff; an examination of the basin showed that it was full of glacial lakes surrounded by dense rain forest (Plate 7.2), where evaporation including interception could approximate to open water evaporation, so that the reservoir would not have resulted in increased basin evaporation. These examples underline the fact that careful interpretation and incorporation of local knowledge are vital when using analytical techniques to understand hydrological regimes.

Comparisons of rainfall and runoff form a common means of extending flow records using long-term rainfall records in the area. The process of correlation which forms part of such analysis (see Chapter 9) will often provide an opportunity of testing the consistency of flow records. This is particularly useful in humid areas, where the basin evaporation loss is likely to be constant from year to year.

Regional comparisons can also draw attention to stations where flow records appear to be suspect, and where attention to rating curves should therefore be concentrated. An initial appraisal of the water balance of a number of sites in Sri Lanka, where a humid climate over most of the project area made comparison simple, quickly revealed (see Fig. 8.2) that certain stations were inconsistent with the others in the region.

Cross-correlation matrices for groups of stations can also identify those stations which are inconsistent with other stations; this may be because the regimes of some basins are different from others, or because the records are not compatible. Knowledge of the hydrology of the area should reveal which of these hypotheses is the more likely.

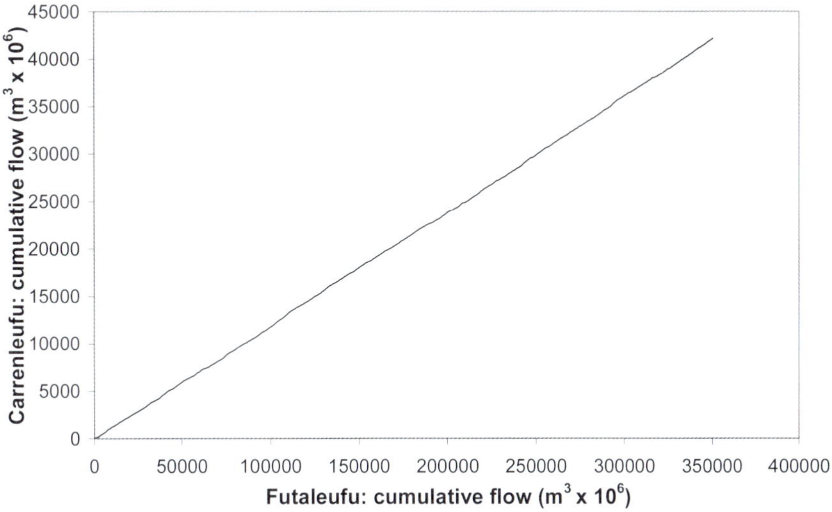

Fig. 7.11 *Comparison of cumulative flows of Futaleufu and Carrenleufu, Patagonia, 1954–1994.*

Plate 7.2 *Rainforest in Futaleufu basin, Patagonia.*

IMPORTANCE OF QUALITY APPRAISAL

It is essential that such checks of record quality should be carried out before time-consuming analyses of the river flow records of a region are embarked upon. Time spent on visiting key stations and examining gauging records and rating curves is seldom wasted, and can prevent a large amount of time being spent subsequently on investigating why certain sites and rivers behave differently from other sites in the region. During the initial investigation of records for the *Flood Studies Report* (NERC, 1975), visits were made to all gauging stations with representatives of the gauging authorities. Assessment of the high flow rating curves and other factors led to quality classification of each flood record.

Examples of flow gauging appraisal

One example where the accuracy of flow records dominated a project review was the Lesotho Highlands Water Project. This project was designed to divert water from the upper tributaries of the Senqu in Lesotho, where the mountainous headwaters provided significant runoff, towards the industrial complex of the Transvaal in South Africa. At the same time hydroelectric power could be generated at the dams through which diversion would occur. The economic value of the project depended on precise estimates of the flows available at key locations, but various estimates of these flows differed significantly. The accuracy of level records, of discharge measurements and rating curves were all examined, and the resulting flow records were subjected to tests of double-mass comparison, water balance derivation and statistical testing. The mean annual precipitation over the project area was closely related to topography, and varied greatly from 600 to 1600 mm as a result of the range of elevation. Consequently the water balance testing was difficult, so that the examination of individual gauging stations gained added importance. Once this quality control had been carried out, it was possible to extend the flow records using long-term basin rainfall records and a deterministic model to provide consistent flow series for system analysis.

The value of the inspection of gauging stations and of gauging methods may be unexpected. During a review of one hydroelectric project in central India, inspection showed that the flow records were based on gauging methods which could best be described as suitable for reconnaissance, though fortunately water balance checks showed that the results were not unreasonable. As part of the review of a second hydroelectric project, the first step was a visit to the key gauging station (Plate 7.1); an invitation to look at a nearby site on an upstream tributary was accepted. This turned out to be a former small irrigation offtake (Plate 7.3) which was the instrument of rapid river capture. The site had been gauged regularly over some 14 years; during this period the flow out of the main river into the channel (Fig. 7.12) had increased from 12% to 24% of the annual flow. Although the comparison was complicated by the fact that the percentage of outflow was higher in years of high flows, the outflow to an adjacent basin had doubled in less than 15 years. Thus a significant proportion of the resource was being lost to river capture at a remarkable rate. This information was taken into account in subsequent planning. It is essential to listen carefully to those with local knowledge and to follow up hints of unusual situations.

Plate 7.3 *Central India: Jawra nullah looking upstream to River Indravati offtake.*

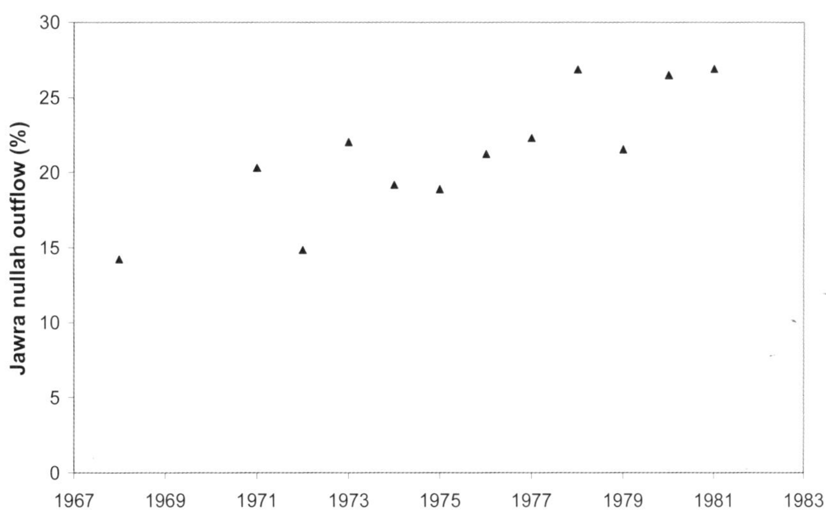

Fig. 7.12 *Central India: Jawra nullah outflow as % of downstream Indravati flow.*

CONCLUSION

The network of gauging stations provides the primary evidence of the water resources of a country. The provision of accurate rating curves should ensure the validity of these records, and therefore the maintenance of a programme of discharge measurements and the monitoring of rating curves is vital to the knowledge and development of a country's water resources. Once the general validity of rating curves and flow records has been established, it is possible to analyse the hydrological behaviour of a region with some confidence. It is therefore most unfortunate that there has been a worldwide reduction in hydrological measurements at a time when the adequacy and stability of water resources come under increased scrutiny. The number of river gauging stations has been decreasing in many countries, and the number of discharge measurements being made to support these stations has fallen even more dramatically. This has occurred at a time when the availability of computer programs for routine analysis of gaugings should have made it easier for local hydrologists to improve the quality of flow records. It is equally important for the quality of records that the local gauging authorities should have the opportunity and resources to carry out water balance and other analyses on the flow records they hold.

8 WATER BALANCE

Once the measurements of the components of the hydrological cycle are complete, the analysis of the hydrology of a basin or a region is best approached through the water balance. This simply expresses the fact that, over any time period, water input to an area must equal output and changes in storage. Input is likely to be mainly rainfall but may include river inflow or groundwater inflow; output will include evaporation, river outflow and groundwater outflow, while storage changes include increases in soil moisture and groundwater storage. Examples of water balance analysis may take the form of balance sheets, but may also include conceptual models which express physical understanding of the links between different processes. It will be seen from examples that the ease of water balance study depends on the type of climate of the area.

TYPES OF CLIMATE

The first step in examining the climate of an area is to compare typical values of average monthly rainfall and potential transpiration or evaporation. Rainfall averages are usually available in standard meteorological yearbooks, but evaporation measurements or estimates are not so readily available as standard statistics. It may be necessary to derive potential transpiration from standard meteorological records using the Penman technique (Chapter 4). In fact climate data are available for this purpose (FAO, 1993). Examples are compared in Fig. 4.1 of average rainfall and potential transpiration for a humid area, a semiarid area and a seasonal monsoon climate. The humid example is from Nuwara Eliya in the central highlands of Sri Lanka, where the incidence of rainfall from both the northeast and southwest monsoons provides a plentiful supply of moisture in most months, and average rainfall exceeds potential transpiration (Fig. 4.1(a)) to give a surplus throughout the year. In these circumstances water balance studies are relatively simple, as the soil moisture deficit seldom exceeds the root constant (Chapter 5) when the transpiration is assumed to fall below the potential rate. In fact runoff can often be estimated as the difference between rainfall and potential transpiration.

By contrast, the comparison between average rainfall and potential transpiration for Francistown in northeast Botswana (Fig. 4.1(b)) reveals that there is no month with a surplus on average. The relatively small depth of annual runoff, typically 10–50 mm, is caused by short periods of intense rainfall in storms which are often isolated and variable. Comparisons between rainfall and runoff must either be based on a comprehensive network of recording raingauges to compare areal rainfall intensity and soil infiltration rates, or be based on time and areal averages and thus remain imprecise. In these semiarid areas estimates of runoff from measured rainfall tend to be empirical, and the runoff process is sensitive to differences in basin topography and geology because the climate leads to low runoff coefficients.

In wide areas of India and Africa there is seasonal variation between arid and humid conditions, and in these conditions the importance of water resources management is evident. These conditions are illustrated (Fig. 4.1(c)) by monthly rainfall and transpiration records for Bhopal in central India. Although the climate appears to be more complex than the other examples, it is possible to simplify the situation by treating the monsoon period as a period of surplus, and the prolonged dry season as a period of deficit with moisture dependent on soil moisture storage.

HUMID WATER BALANCE

The water balance approach is most easily applied in areas of humid climate, where the transpiration can be assumed to be close to the potential rate. Uncertainty arises when the soil moisture deficit increases to the root constant appropriate to the dominant vegetation, when the soil moisture is not readily available and the transpiration falls below the potential rate. Thus the runoff from a humid basin or region should correspond to rainfall less potential transpiration, not just on an average basis but also from year to year. The example of the Tongariro area in New Zealand has already been mentioned in Chapter 3, and is discussed further in Chapter 9. The difference between rainfall and runoff was uniform, so that the flow series could be extended quite simply.

Sri Lanka

A classic case of a water balance study in a humid environment involved the Mahaweli Ganga basin in central Sri Lanka. The Victoria dam project (Piper *et al.*, 1994) required an appraisal of the reliability of the river flow records of the area, and also an extension of flow records to ungauged sites and a common period. The network of long-term raingauges in this region was excellent, as the development of tea estates during the 19th century had led to an interest in rainfall distribution and about 100 stations were available around the central highlands. An annual isohyetal map (Fig. 8.1) was derived from these records and showed that the average rainfall ranges from over 5000 mm on the westerly slopes of the mountains to less than 2000 mm in areas of rainfall shadow which are exposed to neither monsoon direction. The seasonal distribution of rainfall varies with aspect and thus exposure to the southwest and northeast monsoons, which in turn provide rainfall from May to August and from October to January. Other meteorological processes contribute rainfall during other seasons. Although the seasonal distribution over the area as a whole reflects the relative contributions of the two monsoons, the upper Mahaweli basin which feeds the Victoria dam is reasonably homogeneous and rainfall is sufficient to maintain a surplus throughout the year. Even in the tributaries where rainfall is more seasonal, the soil moisture deficit is unlikely to reach a value which would reduce transpiration significantly.

The Penman approach was used to estimate open water evaporation and potential transpiration from meteorological records at three sites in the project area, and these estimates showed a decrease of annual totals with elevation. Flow records were available at a number of sites in the area. A comparison of mean annual rainfall and runoff (Fig. 8.2) showed, as expected, a reasonably constant difference between basin rainfall and runoff depth, which in turn corresponded with potential transpiration. However, this comparison drew attention to the fact that at some sites the runoff records were low, and the flows at these sites were reviewed.

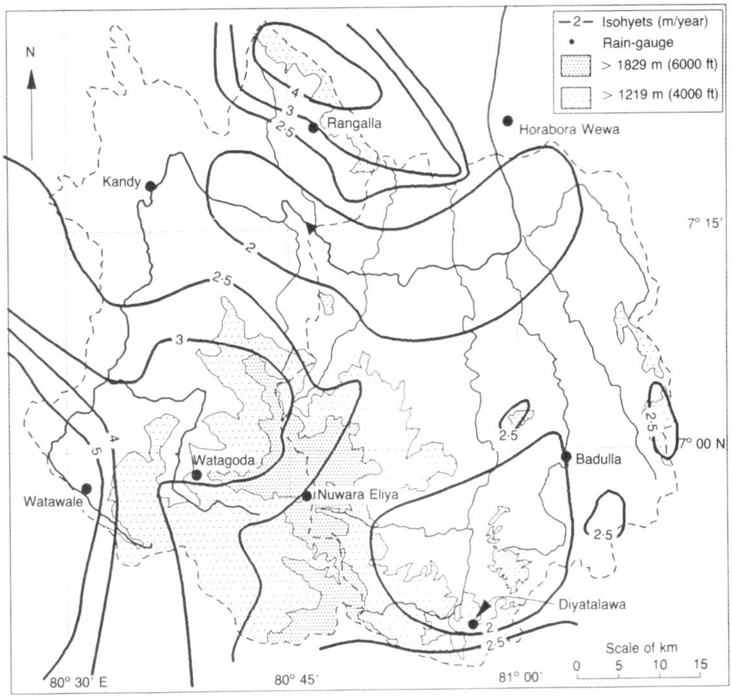

Fig. 8.1 *Central Sri Lanka: annual isohyetal map, m (1931–1960) (after Piper et al., 1994).*

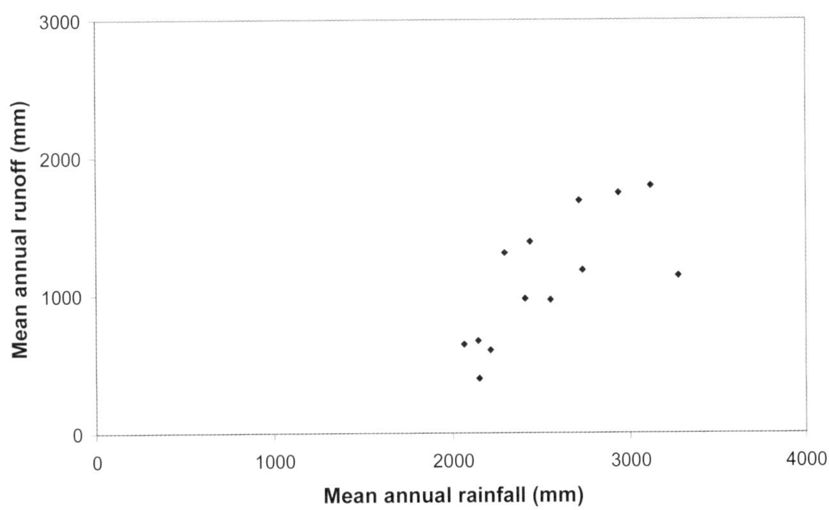

Fig. 8.2 *Sri Lanka: mean annual rainfall and runoff at Mahaweli basin stations (mm).*

There is a similarity between comparisons of mean annual rainfall and runoff at a number of sites in a region, and comparisons of successive annual rainfall totals and runoff depths at a single station. The available records had been checked internally by comparison with Thiessen-based estimates of basin monthly rainfall from selected stations. This information was also used in a comparison of annual rainfall and runoff

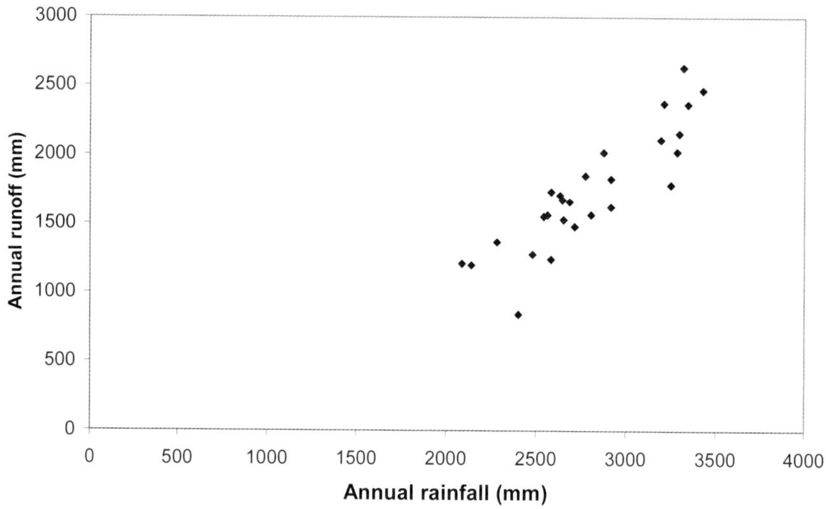

Fig. 8.3 *Sri Lanka: annual rainfall and runoff of the Mahaweli Ganga at Gurudeniya, 1944–1970 (after Piper et al., 1994).*

at all sites, and this comparison confirmed the reliability of most stations. An example (Fig. 8.3) compares the annual flows of the Mahaweli Ganga at Gurudeniya and the basin rainfall series. The correlation between the rainfall and runoff series confirms the reasonable validity of the flow records.

This finding made it possible to estimate mean runoff at key ungauged sites by subtracting potential transpiration from basin rainfall averages derived from the isohyetal map. Consistent series for reservoir operation studies were derived from a statistical comparison between short-term and long-term stations, taking account of cross-correlation and serial correlation within the records. The study was made both simpler and more reliable by the homogeneous and humid nature of the region.

Sarajevo region, Bosnia

In a study in 1998 of the water resources of the Sarajevo region, another example of a humid water balance made it relatively easy to obtain consistent flow records for the area. Sarajevo itself is surrounded by an area of mainly forested hills which are drained by a number of streams which feed into the Bosna River which in turn flows into the Danube. Sarajevo city lies in a valley where a number of these streams converge, and it lies in an area of relatively low rainfall. The monthly average rainfall is compared in Fig. 8.4 with estimates of potential transpiration, derived from Penman estimates of open water evaporation using a factor of 0.8.

It is clear from comparison of Sarajevo rainfall records with those of stations to the north and south that the area receives a combination of Mediterranean winter rainfall with a summer maximum typical of a Central European climate. Although individual years are subject to variations from the average in both sources of rainfall, the average seasonal distribution is extremely uniform. The potential transpiration, on the other hand, is highly seasonal with its link to net radiation and thus temperature. The result is a winter surplus of precipitation, much of which in fact occurs as snowfall, and a relatively small summer deficit.

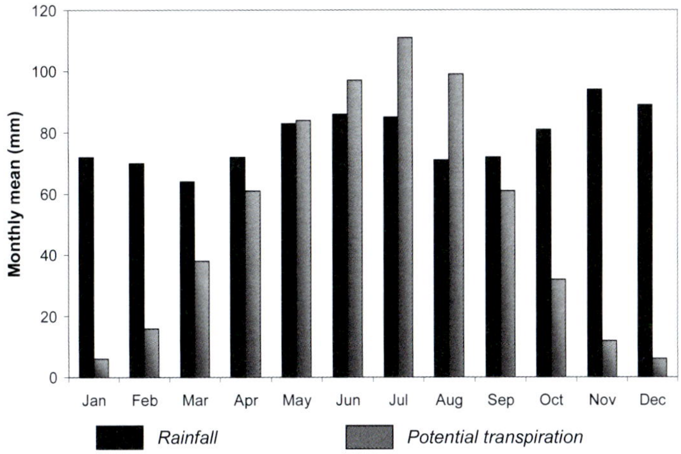

Fig. 8.4 *Sarajevo: mean monthly rainfall and potential transpiration.*

Plate 8.1 *Woodland east (upstream) of Sarajevo.*

It was evident from a visit to the basin headwaters, which enjoy higher rainfall and contribute the bulk of the river flow, that even towards the end of the summer, when the soil moisture deficit would be at its highest, the woodland of the basins (Plate 8.1) appeared to be transpiring freely. It could therefore be assumed that the rainfall was sufficient for the soil moisture deficit rarely if ever to exceed the root constant of the woodland, and therefore that the area could be treated as a humid area. A simple monthly model was developed, with a small fraction (0.1) of the gross precipitation leading to direct runoff, and the balance of the net rainfall contributing to the soil moisture store of 300 mm capacity. This was assumed to transpire at the potential rate and provide runoff when the store becomes full. The runoff was assumed to reach the gauging station over a period of three months; the need for three monthly recession constants to total unity defines them as 0.54, 0.30 and 0.16. An example of this modelling of the River Miljacka at Sarajevo is illustrated in Fig. 8.5. The basin average

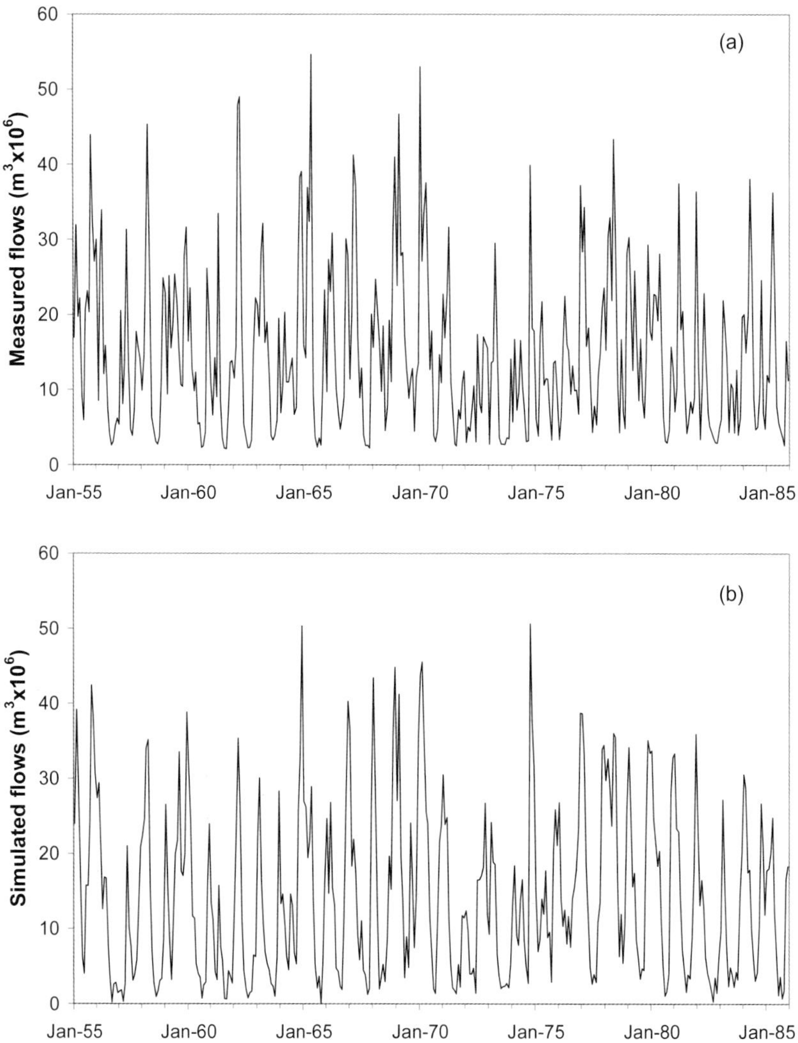

Fig. 8.5 *Miljacka at Sarajevo: (a) measured, and (b) simulated, monthly runoff.*

precipitation was estimated from an isohyetal map as 1200 mm and its monthly distribution from four related rainfall stations using the isopercentile method. This simple model provided reasonable comparisons with measured flows both as monthly series and in a cumulative comparison. The main difference between the measured (Fig. 8.5(a)) and simulated flow series (Fig. 8.5(b)) was a lag in the winter and spring flows which could be explained by snow storage which had not been taken into account in the initial modelling.

This model was then used to generate flow series at ungauged sites, and also to extend all flow series to a standard period in order to represent the flow series over the whole of the contributing area. Again the success of the simple model could be attributed to the nature of the humid water balance.

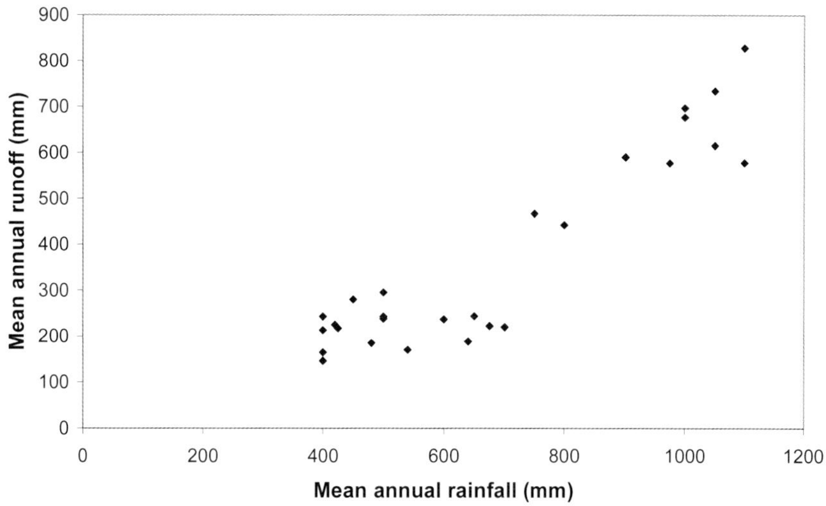

Fig. 8.6 *Northern Turkey: mean annual precipitation and runoff.*

Other humid basins

Areas like upland Britain are evidently humid, as the rainfall in most parts is usually adequate to maintain transpiration at the potential rate through most of the year, and it is rare that the summer has a severe drought during which soil moisture storage cannot maintain transpiration. Most of the runoff occurs during the winter season, but the transpiration occurs at the potential rate throughout the year.

However, there are regions where the maximum rainfall coincides with a season of low potential transpiration. These areas are effectively humid because the runoff derives from a seasonal surplus and the basin loss is relatively constant. For example, the precipitation distribution over the region of northern Turkey inland from the Black Sea is concentrated in the winter months when temperatures and potential transpiration are both low. The result is that runoff occurs between December and April, and is higher than would be expected from the annual precipitation. Figure 8.6 illustrates the mean annual precipitation and runoff for a number of basins. The runoff corresponding to basins of 500 mm precipitation is about 200 mm, while for basins with precipitation of 1000–1100 mm the runoff is as high as 600–700 mm. This is much higher than the runoff in other parts of the world with similar precipitation, for example Botswana and Yemen described in the next section, because of the low potential transpiration during the winter precipitation season.

ARID WATER BALANCE

Water balance analysis is less simple in arid or semiarid areas, mainly because actual transpiration is much less easy to predict from potential transpiration. Because surplus soil moisture seldom gives rise to widespread runoff, the runoff process is more likely to depend on surface flow from local intense storms when rainfall exceeds local infiltration. This is of course the classical Hortonian concept of runoff, based on United States experience, which has not been entirely convincing to hydrologists working in more temperate conditions, who may never have observed surface runoff.

Such conditions require more complex analysis than the humid situation described previously. If the runoff process were to be modelled physically, it would be necessary to monitor rainfall on a very detailed scale, both in terms of the time scale required to estimate rainfall intensity, and also to take account of the greater spatial variability in shorter time intervals. However, it will be shown that it is possible to use models based on monthly time intervals by making assumptions about the distribution of rainfall and basin conditions.

Northeast Botswana

Successive studies in Botswana in 1968, 1976 and 1986 (Parks & Sutcliffe, 1987) illustrate the problems which semiarid climates present. Average annual rainfall decreases from over 500 mm north of Francistown to less than 350 mm in an area of low rainfall in the lower Motloutse basin to the south (Fig. 8.7). The rainfall season is from November to March, when the potential transpiration is at its highest. Monthly averages for Francistown (Fig. 4.1(b)) show that annual rainfall is less than a third of transpiration and its seasonal cycle is similar. Thus potential transpiration exceeds average rainfall in all months, and runoff mainly results from local convective storms and high rainfall intensity. Annual runoff is low at a basin average of 10 to 50 mm.

In fact it was only after 20 years of fairly regular visits that it was possible to observe significant runoff for the first time (compare Plates 8.2 and 8.3). During heavy

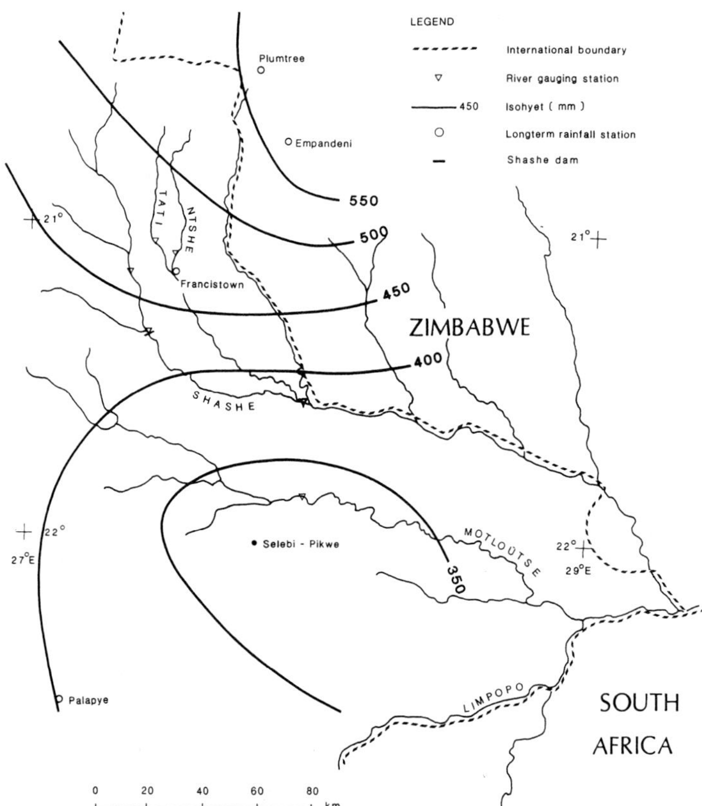

Fig. 8.7 Northeast Botswana: average annual rainfall, mm (1939–1969) (from Parks & Sutcliffe, 1987).

Plate 8.2 *River Shashe gauging weir near Francistown, Botswana.*

Plate 8.3 *Shashe weir during spate, 1986.*

storms runoff appears to occur over parts of the basin where rainfall intensity exceeds infiltration rates on less permeable soils, bringing into operation classical hydrological concepts like depression storage and surface runoff. The runoff generated locally forms sheet flow and is concentrated in gullies and small streams and thus into river flows.

The headwaters of the Shashe tributaries north of Francistown are on Basement Complex rocks with moderate relief and soil depth, and gullies develop into sandy tributaries with rock outcrops and finally into broad sand-bed rivers draining to the Limpopo. To the southwest the Kalahari Beds provide deep sand cover, especially in the headwaters of the Motloutse (Fig. 8.7). River flows occur in a small number of spates, but leave the sandy river beds apparently dry for much of the time, even in the wet season. As a result flow measurement is extremely difficult, but a combination of gauging weirs on rock outcrops (Plate 8.2) in the upper tributaries and rated sections in the lower reaches have over the years provided records at key sites. The number of significant streams is limited, and the locations of gauging sites were partly determined by early investigations of potential dam sites.

In these conditions, water balance studies are likely to be largely empirical, as physical modelling would require a dense network of measurements. It is possible to compare average runoff depth at gauging sites with mean basin rainfall, taking note of other factors such as geology, soils, topography and vegetation. Alternatively statistical or semi-empirical conceptual models may be used to extend flow records where sufficient rainfall and runoff records are available for calibration, or to generate approximate flow series at ungauged sites.

In early investigations in 1968, only three flow records each of six years' duration were available, and a detailed isohyetal map had to be compiled as a basic tool of the study. Use was made of regional studies in southern Africa by Midgley (1952), which suggested that although mean runoff depth depends on a number of factors like rainfall, relief, geology, soils and vegetation, these factors are to a large extent linked and relations between mean rainfall and runoff depth can be determined in terms of regional vegetation and land use. He provided a table linking mean runoff with rainfall in zones classified by vegetation and land use, based in part on flow records on Limpopo tributaries.

The mean runoff at the three gauged sites was deduced by comparing rainfall during the period of flow records with the long-term mean, and the results were compared with Midgley's table. However, the sensitivity of the choice of zone made it risky to estimate the runoff at ungauged sites, and a reservoir site on the Shashe below the gauging station was selected for detailed study. A simple relation between monthly rainfall and runoff was derived and used to deduce the probable reservoir level at the start of the critical drought. This study drew attention to the importance of measured flows in the area, and gauging was extended to other potential reservoir sites.

A later regional reconnaissance study of potential reservoir sites in eastern Botswana was carried out in 1976. After likely dam sites had been identified from air photography, estimates of mean annual runoff and its variability were required at 45 sites in northeast Botswana. Flow records at eight sites in northeast Botswana were adjusted to a common period and runoff depths were compared with mean basin rainfall, together with 10 records from adjacent Shashe tributaries in Zimbabwe. On the assumption that land use and slopes were reasonably homogeneous, a relationship

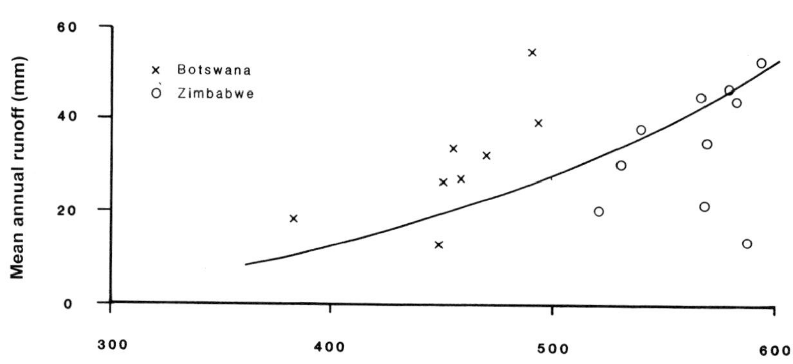

Fig. 8.8 Northeast Botswana: mean annual rainfall and runoff (mm) (from Parks & Sutcliffe, 1987).

between runoff and rainfall (Fig. 8.8) was used to make preliminary estimates of mean runoff at reservoir sites, from which a smaller number were selected for further study. The variability of annual flows was deduced from dimensionless cumulative flow frequency curves for durations from 1 to 7 years derived from adjacent South African flow records (Midgley & Pitman, 1969). These were tested using a rainfall–runoff model of the Shashe flows.

By 1986, when a third study was made, these and other studies had resulted in the installation of gauging stations near the most promising reservoir sites. Measured flows were available for periods ranging from 15 to 23 years. This period was thought sufficient to fit a rainfall–runoff model to extend the flows over the period of long-term rainfall records. Rainfall records at Francistown and at two other sites with records from 1922, were supplemented by adjacent stations in Zimbabwe, some of which dated from 1899. The isopercentile method (Chapter 3) was used with the basin averages to derive long-term rainfall series for each gauged basin.

Pitman model

The Pitman model (1973) was fitted to the observed rainfall and runoff series for each basin. This model (Fig. 8.9) was developed to represent hydrological processes over a variety of climates in South Africa, and was initially designed to convert monthly rainfall depths and potential transpiration data to monthly river flows. It is semi-empirical in that the rainfall–runoff process is simplified and expressed in semi-statistical relations. The model takes into account interception storage, the impervious fraction and a range of infiltration rates, soil moisture storage and basin lag. The model uses 12 parameters, some of which are important in humid areas and others in semi-arid areas. Monthly rainfall is converted to a synthetic symmetrical S-shaped curve, in which the deviation from a uniform rainfall rate is related to monthly totals. Monthly interception is deduced from monthly rainfall as a function of interception storage. Runoff in semiarid areas is related to the impervious fraction and the variation of infiltration rates over the basin, and is modelled by a symmetrical triangular frequency distribution between the minimum and maximum infiltration rates.

Where runoff depends on water balance rather than infiltration rates, basin evaporation is assumed to decrease below the potential rate as soil moisture storage decreases below the maximum, while the runoff coefficient is linked to soil moisture by a power relation with a threshold soil moisture and a runoff coefficient at maximum storage. Runoff is lagged by treating the basin as a linear reservoir. Although the full model has 12 parameters, only five were important in these semiarid conditions: the maximum soil moisture capacity, the minimum and maximum catchment absorption rate, the interception storage, and the runoff lag.

The fixing of the model parameters is based on comparison of simulated and observed flows. Although several of the parameters have hydrological titles, it is recommended (Pitman, 1973) that their estimation should be adjusted by fitting the modelled to the observed runoff. The calibration of the model was based on comparisons of simulated and observed mean annual flow, the mean and standard deviation of the logarithms of annual flows, and the seasonal distribution of runoff. Advice is given by Pitman on the effect of changes in the model parameters; once the parameters had been adjusted by recommended methods, the Botswana runoff series were extended back to 1922 using long-term rainfall series. However, the lack of close correlation

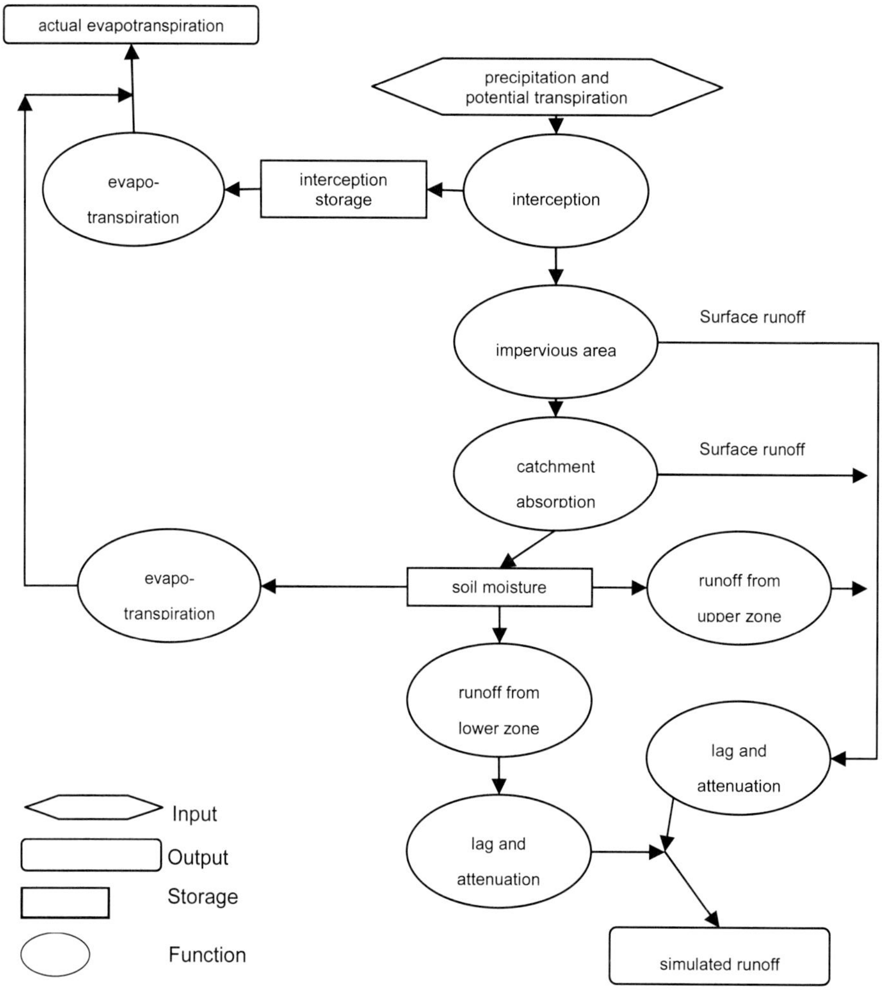

Fig. 8.9 *Flow diagram of the Pitman model (after Pitman, 1973).*

between simulated and observed runoff during the calibration period (Fig. 8.10) shows that the natural variability of the runoff process (Plate 8.3) in this semiarid environment makes it difficult to model the rainfall–runoff process accurately in this area.

As the available flow records for this region have increased in number and length, they have enabled more comprehensive methods of analysis to be used. Once a reasonable period of runoff records could be used, it was possible to use rainfall records to extend this period further. As a result of the increasing length of direct or extended flow record, the standard error of estimate of the mean annual runoff at the Shashe dam has decreased steadily, from 30 $m^3 \times 10^6$ in 1968 to 20 $m^3 \times 10^6$ in 1976 and 12 $m^3 \times 10^6$ in 1986. However, the area appears to be subject to fluctuating periods of dry and wet years (Fig. 8.11), and the successive estimates have varied from 86 $m^3 \times 10^6$, up to 99 $m^3 \times 10^6$ and down to 74 $m^3 \times 10^6$ over the years. In view of this variability of available water resources which occurs over the years in some climates, it is important to use as long a period of flow record as possible. This topic is discussed further in Chapter 9.

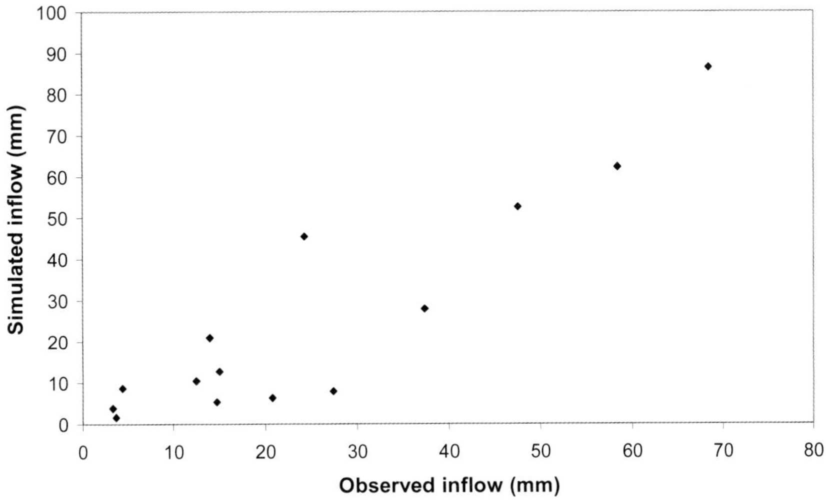

Fig. 8.10 *Shashe, northeast Botswana: measured and simulated annual runoff, 1970–1983 (after Sutcliffe, 1995).*

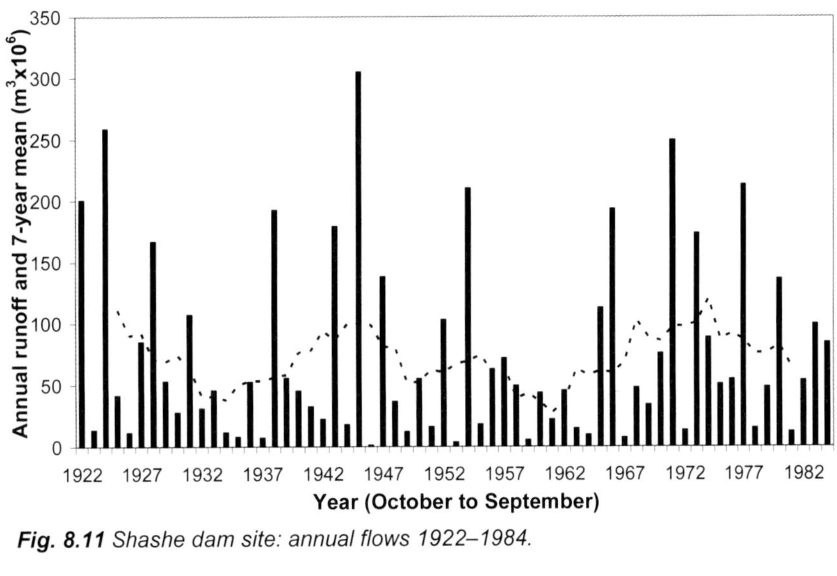

Fig. 8.11 *Shashe dam site: annual flows 1922–1984.*

Iran

Two examples from Iran illustrate the problems of deriving a water balance in mountainous areas. The first example, from the Zagros Mountains in western Iran (Sutcliffe & Carpenter, 1967), is mentioned in Chapter 3 as the estimation of mean precipitation over the different river basins required adjustment to a common elevation (Fig. 3.5) as precipitation varied with both elevation and position. Although the mean precipitation varied between 300 and 700 mm, the mean runoff depth varied between 40 and 300 mm. The open water evaporation, estimated by the Penman method, decreased from 2000 mm at an elevation of 150 m at Dezful, to 1600 mm at 1300 m at Kermanshah, but the seasonal distribution of precipitation was such that the surplus

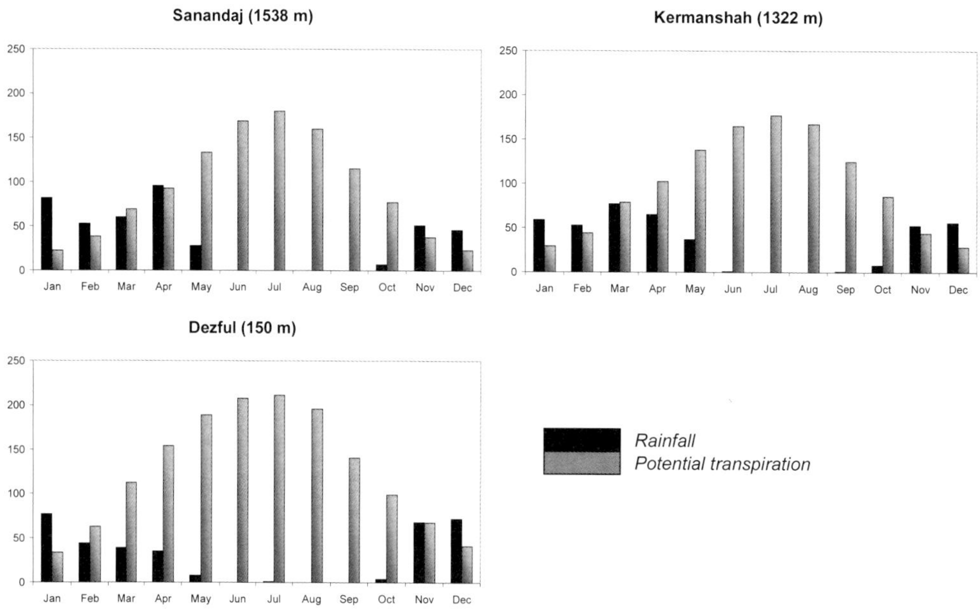

Fig. 8.12 *Kermanshah area, Iran: rainfall and potential transpiration at key sites, mm.*

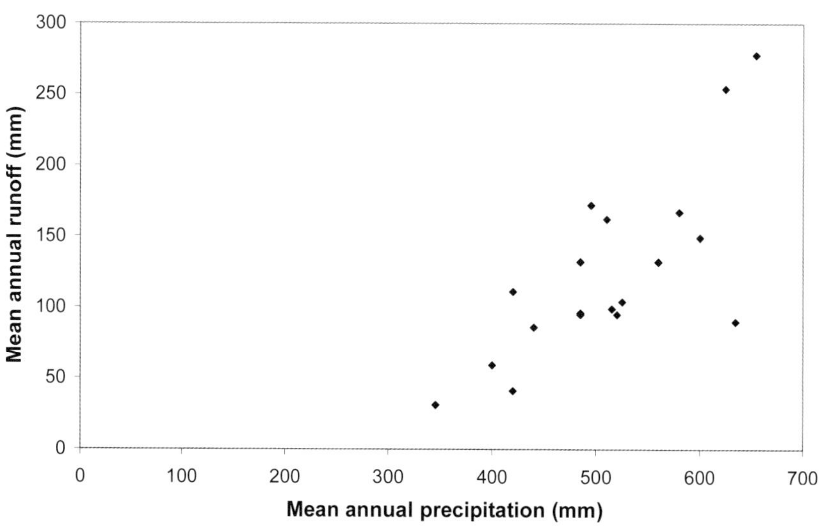

Fig. 8.13 *Zagros region, Iran: rainfall and runoff (after Sutcliffe & Carpenter, 1967).*

was relatively large. The seasonal distributions of precipitation and potential transpiration at key sites are illustrated in Fig. 8.12. These show that a small water surplus occurs when the winter precipitation exceeds the low transpiration, particularly at high elevations. On the other hand, the potential transpiration is high during the summer, when there is little or no rainfall. The relation between mean precipitation and runoff at sites where measured flows were available, after adjustment to a common period, is illustrated in Fig. 8.13. Further analysis of the records revealed that actual evaporation

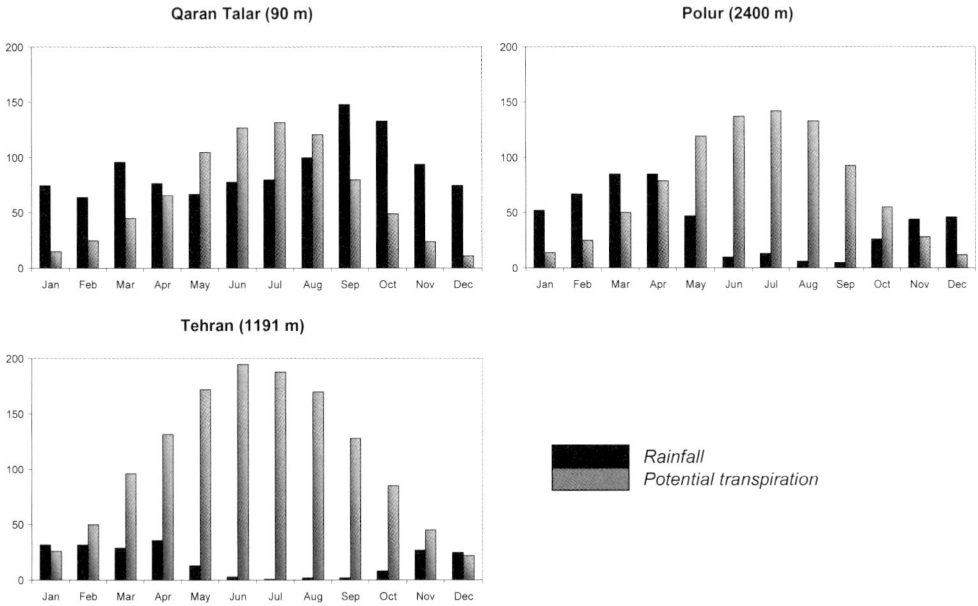

Fig. 8.14 *Alborz area, Iran: rainfall and potential transpiration at key sites, mm.*

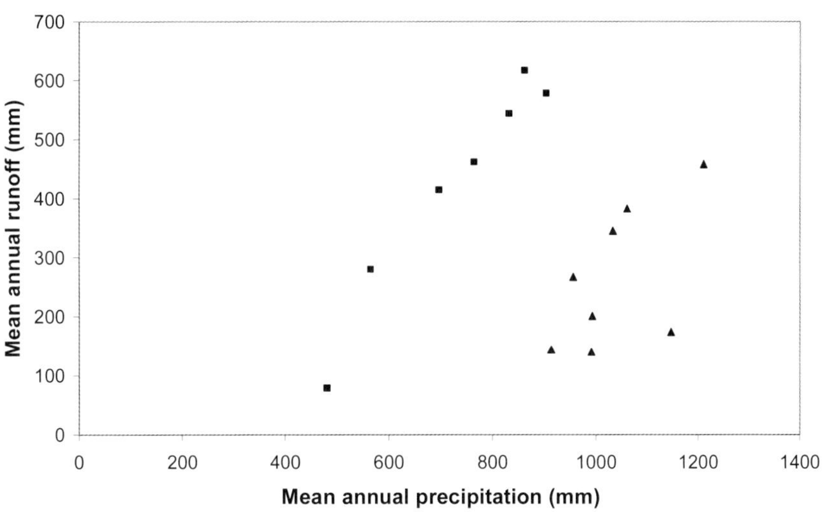

Fig. 8.15 *Alborz region, Iran: rainfall and runoff; ▲ northern and ■ southern basins.*

is quite conservative. Indeed, though precipitation depends greatly on elevation, it appears that the combined effect of lower potential transpiration and higher precipitation at higher altitudes produce a fairly constant actual evaporation loss. This fact made it possible to make reasonable estimates of mean runoff at ungauged sites.

A second example from Iran contrasts a humid region with an adjacent semiarid region separated solely by aspect (Sutcliffe & Swan, 1970). To the south of the Alborz massif, on either side of Tehran, are a number of basins with a water balance similar to the Zagros region in western Iran. The mean precipitation increases from 210 mm at Tehran at an elevation of 1200 m, to some 800 mm at elevations of 2500 m on the southern slopes, but the precipitation is confined to the winter months. To the north of the watershed, on the basins facing the Caspian Sea, the precipitation is higher, up to 1200 mm, with incidence throughout the year. Monthly average precipitation and potential transpiration at key stations to the north and south of the divide are compared in Fig. 8.14, which illustrates the contrast between the climate of the two aspects. This contrast was discussed in Chapter 3 and the isohyetal map was shown as Fig. 3.6. However, the even distribution of rainfall on the northern slopes has allowed forest to survive and become thick in some areas, and the measured runoff from the northern rivers is in fact slightly lower than from the drier but relatively bare basins to the south.

The mean runoff values from basins on either side of the mountains are compared in Fig. 8.15; the difference between precipitation and runoff, or the apparent loss, is about 750 mm on the northern slopes but only about 300 mm on the southern slopes. The difference in climate between the two slopes is responsible for the difference in vegetation, but the vegetation in turn is responsible for the higher transpiration and interception loss on the forested basins. Once the concept of soil moisture storage is included in the comparison of precipitation and transpiration, the effects of the different climates on runoff depth are explicable. This area provides an interesting comparison of a humid and a semiarid set of river basins separated simply by the Alborz watershed.

Yemen

A more extreme example of a water balance in an arid climate was provided by an investigation in Yemen (Farquharson *et al.*, 1996). The topography of Yemen (Fig. 8.16) is dominated by mountain ranges running parallel to the Red Sea coast and rising to over 3700 m. The rainfall is affected by the topography and varies from about 800 mm to the west of the main mountain range to less than 50 mm along the Red Sea and Gulf of Aden coastal plains, and also inland towards the *Rub el Khali* or "Empty Quarter". The rainfall is confined to two seasons, around March–May and July–September, and the relative importance of these two seasons varies with proximity to the two coasts. Analysis revealed that the annual total rainfall varied with the number of storms rather than their intensity, and storm magnitudes were reasonably uniform over the region. With annual potential transpiration over the area at about 1800–2000 mm, the runoff is extremely intermittent and depends on the incidence of storms. An unusual feature of the land use (Plate 8.4) is that much of the area is subject to bunding and terracing, even on steep hillsides, to detain runoff and permit rainfed agriculture.

Long-term measurements of runoff were rare, but a combination of rainfall modelling, based on the typical number of storms, and runoff modelling based on classification of land types and antecedent conditions, was used to extend several flow records. The seasonal distribution of runoff from the longer records reflected the variation over the country of seasonal rainfall. The inclusion of shorter records in a comparison of mean annual rainfall and runoff depth (Fig. 8.17) showed a fair degree of scatter but no differentiation between different regions. The comparison suggested that the runoff coefficient does not drop below 5% even for the most arid basins. This

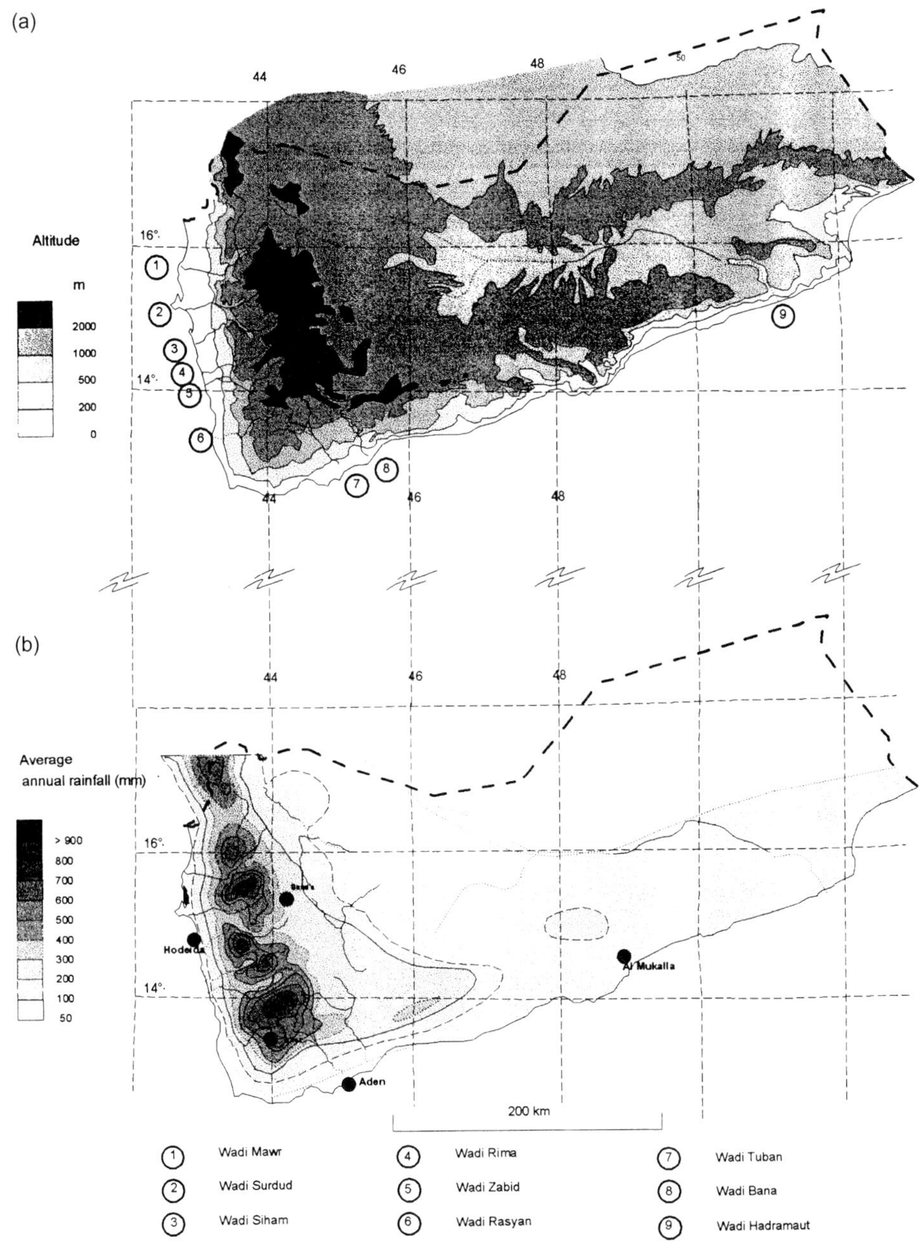

Fig. 8.16 *Yemen: (a) topography and (b) rainfall (from Farquharson et al., 1996).*

does not accord with the expectation that the runoff coefficient tends to zero as the rainfall decreases, but may be consistent with the rainfall model. If the storm magnitudes are not related to annual average rainfall, and antecedent rainfall is low, the runoff generated by each storm will depend on basin characteristics such as geology, which are constant.

Plate 8.4 *Terracing west of Sana'a, Yemen.*

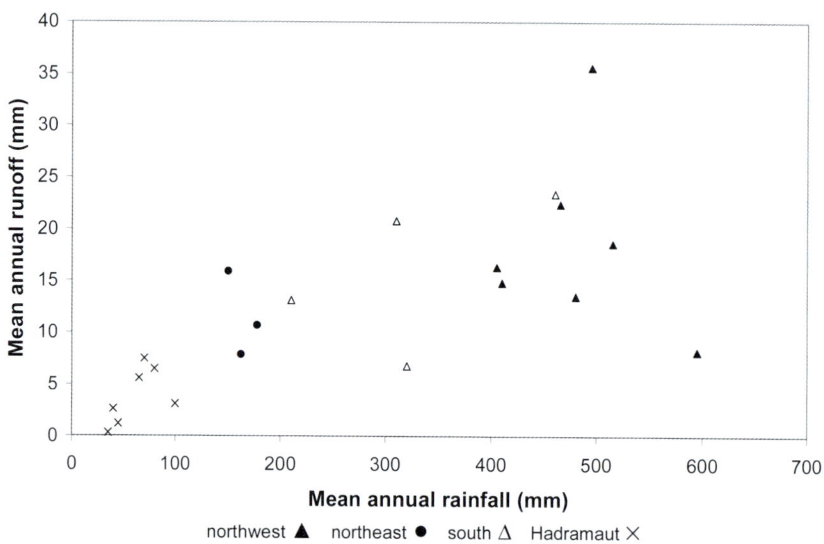

Fig. 8.17 *Yemen: rainfall and runoff (after Farquharson et al., 1996).*

MONSOON CLIMATE

Betwa basin

The water balance of the Betwa basin in central India has been discussed in Chapter 5. In this basin a relatively short rainfall period is followed by a long dry season when the only available moisture is provided by soil moisture storage. Part of the net rainfall during the monsoon season is retained as soil moisture and maintains transpiration during the dry season until the storage is exhausted. The way in which the depth of the soil moisture store can be deduced from the balance between a 50-year measured series of seasonal net rainfall and runoff depth was described in Chapter 5, and a direct method of measuring groundwater recharge was described in Chapter 6. Further studies (Sutcliffe & Green, 1986) were carried out on a tributary of the main river. Within the Nion basin in the centre of the Betwa basin to the east of Vidisha (see Fig. 3.8), detailed measurements of the different components of the water balance were taken over two monsoon seasons. Fortuitously, these two seasons were very different; one was slightly above average in terms of rainfall, while the second was an exceptionally dry year. This investigation was followed by the formulation of a simple hydrological model which expressed the understanding gained of the rainfall–runoff process.

A gauging station was established in 1977 on the Nion about 10 km above the confluence with the Betwa, with a contributing area of 921 km^2; rating was by current meter. Additional rainfall stations were installed to provide six daily records, and basin averages were calculated by the isopercentile method. Daily potential transpiration estimates were derived from meteorological records at Bhopal using a variant of the Penman method. A network of eight soil moisture sites (Hodnett & Bell, 1986) was monitored at frequent intervals from September 1977 to September 1978 as part of the wider programme, with five in wheat, two in pulses and one in rough grass, in proportion to actual land use. In 1978–1979, when rainfall was exceptionally low, a weighted mean was deduced from just four sites. The basin storage was estimated from the mean storage down to a depth of 2.1 m. During the 1978 monsoon, groundwater levels were measured at five observation wells in the basin, and in 1979 were derived by correlation from the one well still operating; the changes in well level were converted to storage change using a storage coefficient estimated from pumping tests to average 0.1. These estimates of basin rainfall, transpiration, runoff, soil moisture storage and groundwater storage are summarised as mm over the basin in Fig. 8.18(a) and (b) for the monsoon months in the years 1978 and 1979. The timing of the rainfall input and its dispersion through storage and runoff are also illustrated clearly by the cumulative analysis of Fig. 8.19. Because the rainfall in the 1978 monsoon, at 1235 mm, was above average, while the 1979 rainfall, at 460 mm, was the lowest on record, the contrast between the two years throws light on the water balance processes. It is clear from the cumulative depths in the different stores that rainfall initially recharges the soil moisture store, and not until this is full does any net rainfall reach groundwater or surface runoff. Once the groundwater store is full, when the water level in this area reaches the surface, all the net rainfall becomes runoff.

In the first phase of the rainfall season of 1978 (days 1–25), rainfall goes almost entirely to soil moisture recharge. In the second phase (26–40), soil moisture recharge continues but the shallow aquifer also responds to groundwater recharge and some runoff occurs. During the third phase (41–98) there is little further change in soil moisture or shallow aquifer storage, and the bulk of the net rainfall becomes runoff; most of the runoff occurs during this phase. In the fourth phase baseflow runoff is maintained by groundwater storage, and actual evaporation from the basin depends on soil moisture and the occasional shower. In fact the recession curve of baseflow can give information directly about groundwater storage; this is particularly useful in a highly seasonal environment, as the volume of runoff after the end of the rainfall season can easily be estimated from the hydrograph and converted to storage depth over the basin.

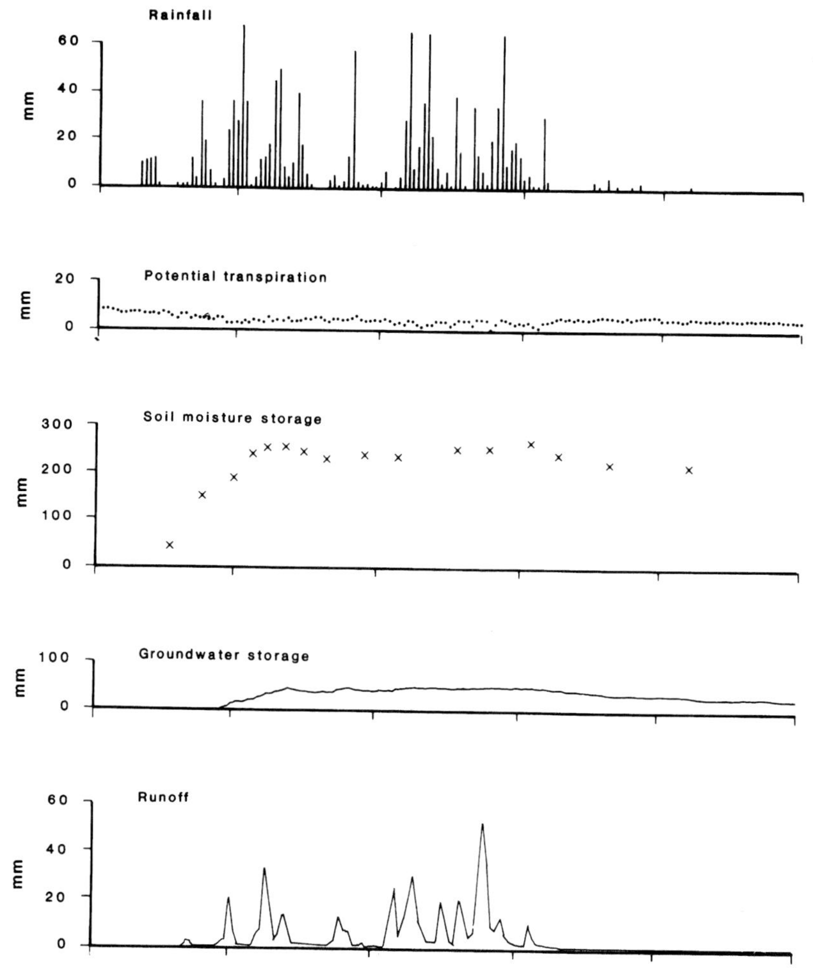

Fig. 8.18(a) *Nion basin, central India: water balance in 1978 (from Sutcliffe & Green, 1986).*

The water balance for 1979 was very different because of the very low monsoon rainfall. However, the sequence of events was similar, and it is the later processes in this sequence, like groundwater recharge and runoff, which were most affected by the low rainfall. Figure 8.19 shows that soil moisture recharge was little lower than in 1978, while the transpiration during the monsoon was below the potential rate. The groundwater recharge occurred later in the season, but was similar in total to that in 1978. Although the runoff depth was severely reduced, the depth of rainfall which occurred before runoff was generated was similar to the previous very different year.

Conceptual model

A simple conceptual model, designed to reproduce the links between the processes, is illustrated in Fig. 8.20. This model was developed from a basic understanding of the physical processes which were dominant within the Nion basin; because this basin was

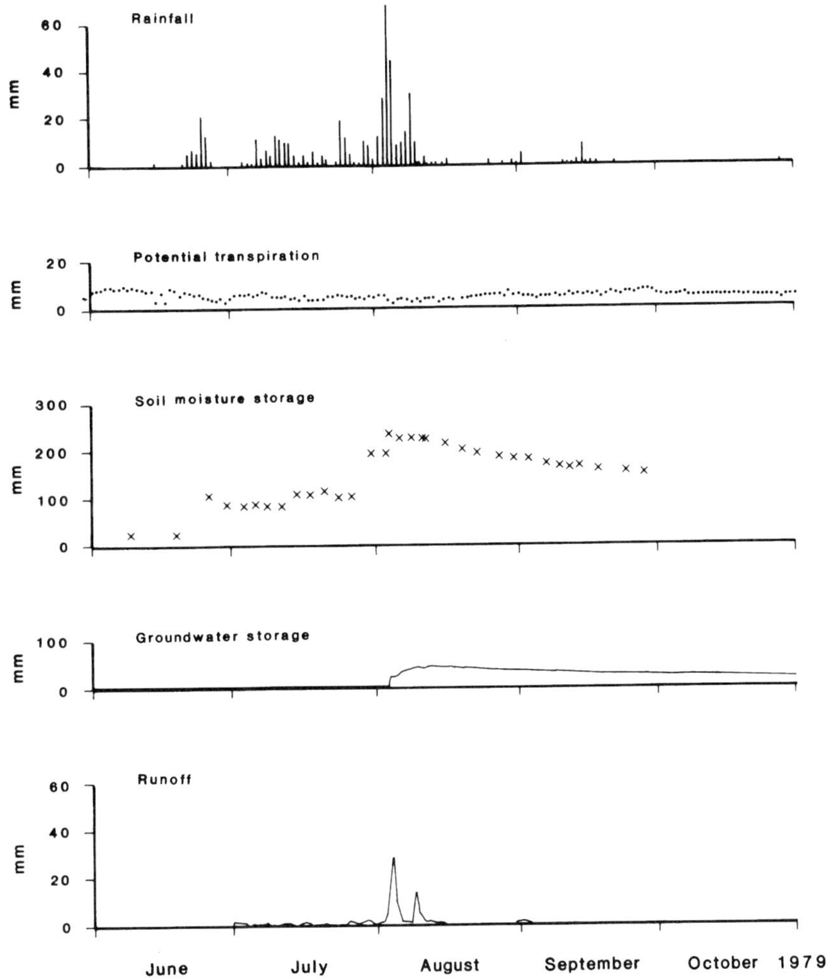

Fig. 8.18(b) *Nion basin, central India: water balance in 1979 (from Sutcliffe & Green, 1986).*

remarkably homogeneous in terms of soils and geology, it was possible to represent it by a lumped model, in which the basin is represented by a single point. The "black cotton soils" overlying weathered basalt provided an uniform soil moisture store which largely controlled the partition of rainfall surplus first into soil moisture storage, then into groundwater and surface runoff. The model was able to represent these successive processes as single storages, which were filled in turn. The model (Fig. 8.20) consisted of a single soil moisture store, a groundwater store, and a channel store. The rainfall, R, is added to the soil moisture store, S; when this exceeds a first threshold, T_1, evaporation occurs at a rate a_1E, proportional to potential transpiration, and drainage, $a_2(S - T_1)$, to groundwater and runoff is proportional to this excess storage. When the soil moisture, S, exceeds a second threshold, T_2, then all the excess drains to groundwater, and to runoff in a proportion fixed by a parameter, a_4, between 0 and 1. When there is water, G, available in the groundwater store, drainage to runoff, a_3, is proportional to this store; when G exceeds a threshold, T_3, the excess runs off directly.

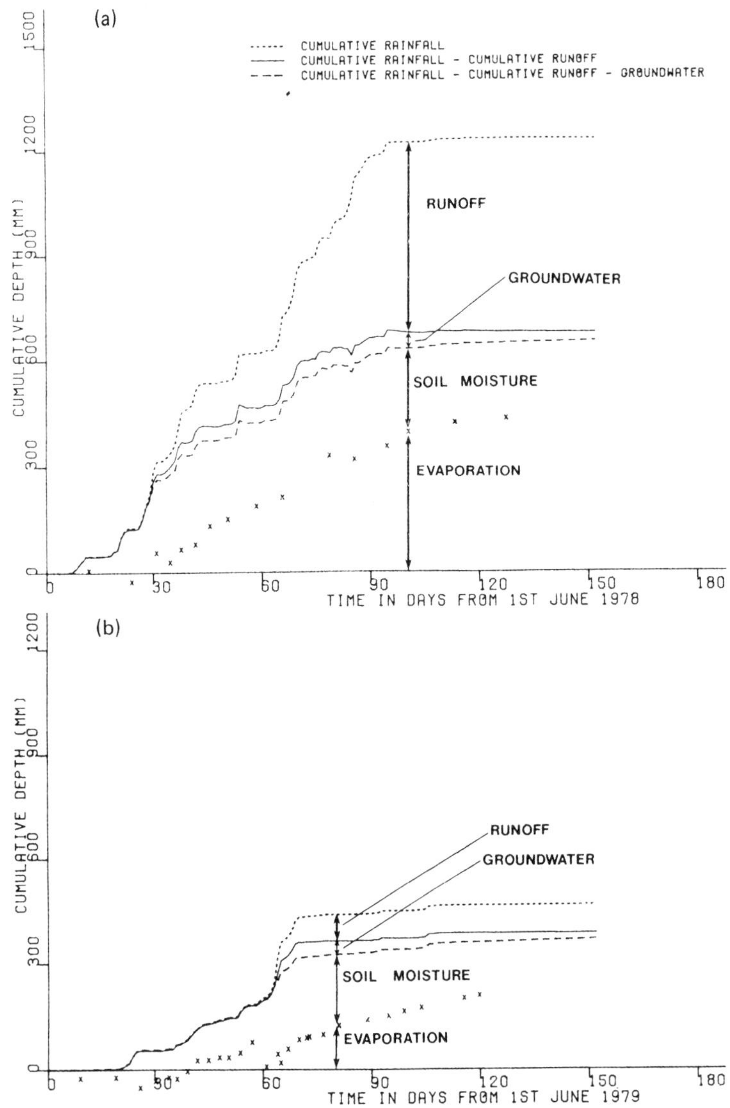

Fig. 8.19 *Nion basin, central India: cumulative water balance in 1978 and 1979: (from Sutcliffe & Green, 1986).*

The runoff is routed through a linear reservoir so that $Q = a_5V$. There was also a small allowance for losses before the monsoon, and the evaporation was reduced during the fallow period at the end of the monsoon.

Table 8.1 *Estimated values of model parameters.*

a_1	1.0	T_1	150 mm
a_2	0.0025	T_2	250 mm
a_3	0.014	T_3	50 mm
a_4	0.65		
a_5	1.8		

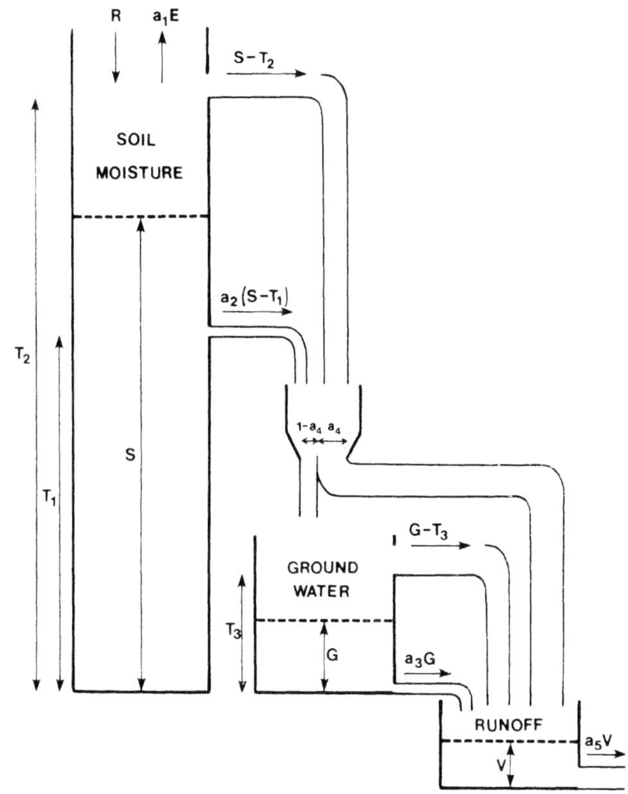

Fig. 8.20 *Nion basin, central India: conceptual model (from Sutcliffe & Green, 1986).*

Although this model has eight parameters, a_1–a_5 and T_1–T_3, it is simple and the parameters have physical meanings. Because direct measurements of soil moisture and groundwater storage were available, some parameters (Table 8.1) were estimated directly and others were adjusted by comparing predictions with measurements. The predicted and measured values of soil moisture storage, groundwater storage and runoff, were found to correspond well in two successive seasons with strongly contrasting rainfall inputs. Comparisons between predicted and observed values in 1978 (Fig. 8.21) suggest that a reasonable representation of the physical processes was achieved. The model was very simple in concept, but was also used elsewhere as a forecasting tool.

Ancient water resources development

The relative simplicity of the water balance in the monsoon climate of the Betwa basin, and the evident benefits of ensuring the availability of water throughout the long dry season, doubtless led to the construction of a series of reservoirs which have been dated to the period starting about 150 BC in this area. During an archaeological survey of an area within about a 15 km radius of Sanchi hill, near Vidisha, a centre of Buddhist propagation in central India, a set of 16 ancient dams was discovered (Shaw, 2000); these were almost all dated to between the 2nd–1st centuries BC and the 5th century AD, using *naga* serpent sculptures associated with the structures, stone facing and associated archaeological sites. Simple estimates of the reservoir volumes associated

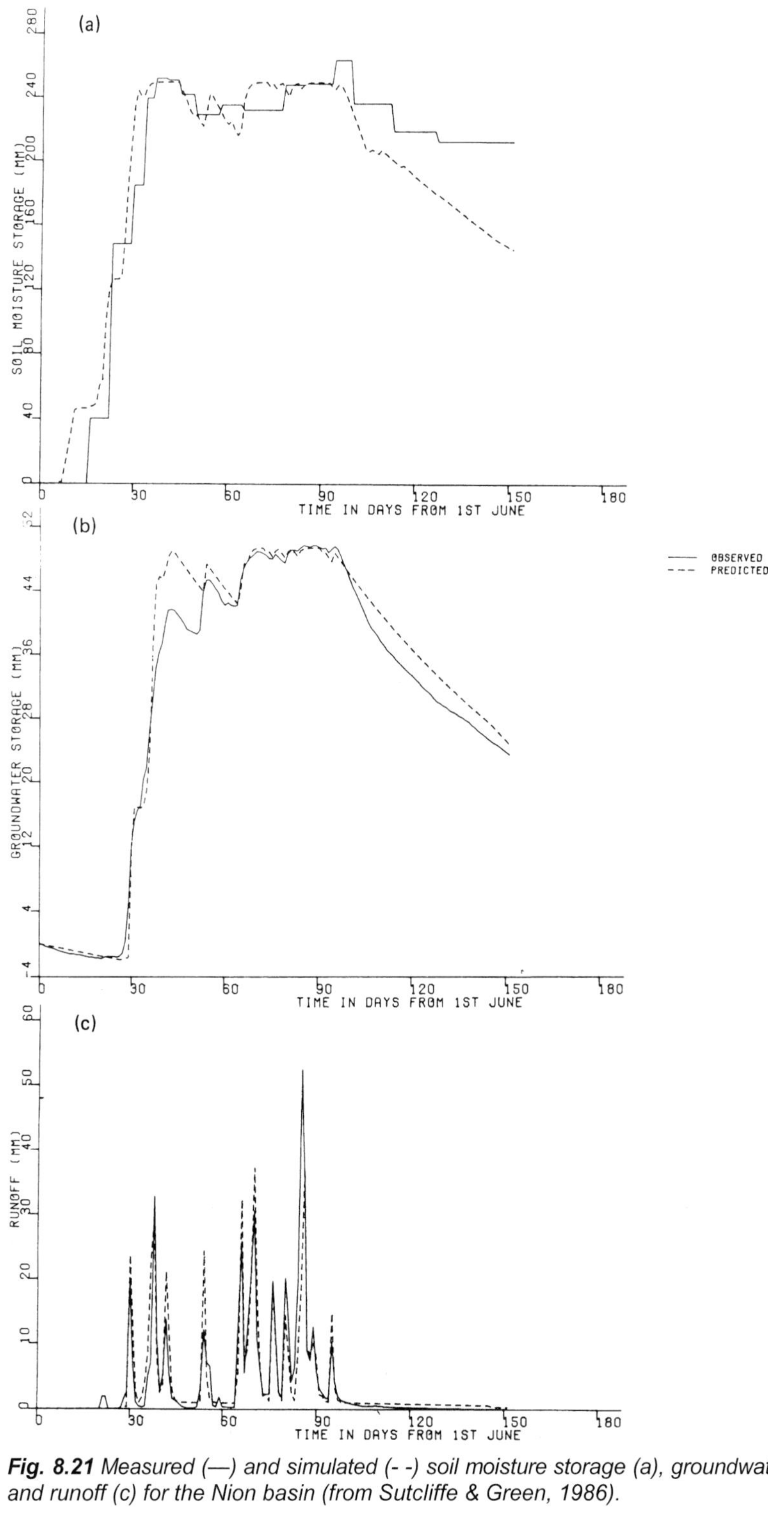

Fig. 8.21 *Measured (—) and simulated (- -) soil moisture storage (a), groundwater (b), and runoff (c) for the Nion basin (from Sutcliffe & Green, 1986).*

Plate 8.5 *Masonry on the upstream side of the Morel Kala dam, central India.*

Plate 8.6 *Spillway on the Devrajpur dam, central India (from Shaw & Sutcliffe, 2001).*

with these dams were made from measurements of dam heights, lengths and reservoir gradients. When these were compared with runoff volumes derived from basin areas and runoff depths from the Betwa study, it was found that the reservoir volumes were closely related to the runoff volumes (Fig. 8.22), which suggested that the dams were designed with an understanding of the principles of water balance. It is suggested (Shaw & Sutcliffe, 2001) that the reservoirs were designed to irrigate rice for the increased population related to the Buddhist expansion in the Sanchi area. Most of the dams are protected on the upstream face with well-constructed masonry (Plate 8.5) and at least two of the later dams are also protected by spillways (Plate 8.6) which appear to have been designed to pass the 50-year flood (Shaw & Sutcliffe, 2003).

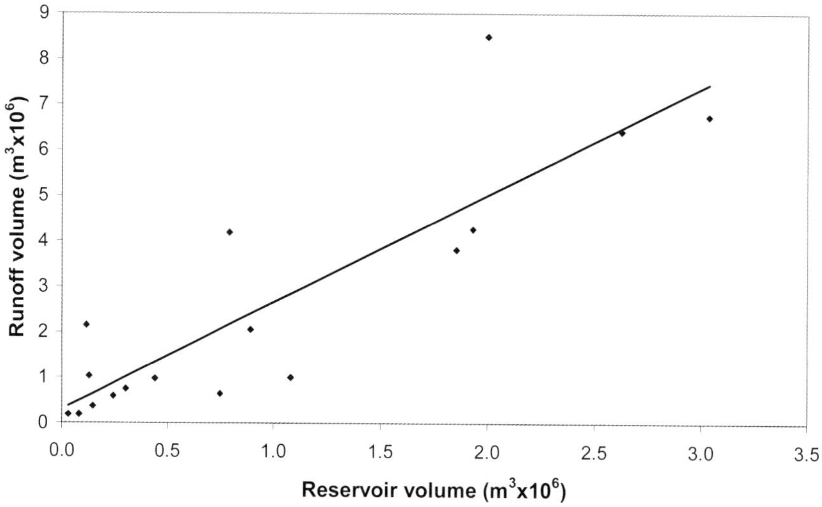

Fig. 8.22 *Ancient dams near Sanchi: reservoir volume versus runoff volume (after Shaw & Sutcliffe, 2001).*

West Africa

The methodological findings of the Betwa study were applied to a regional investigation of water resources in Guinea and Togo-Benin in West Africa (Sutcliffe & Piper, 1986), where the seasonal rainfall distributions were similar to Betwa and the deciduous woodland suggested that the soil moisture regimes were similar. In reviews of potential hydroelectric power resources, runoff estimates were needed at a number of potential reservoir sites over the region.

From monthly series of rainfall and potential transpiration, seasonal net rainfall series for basins were derived and compared with annual runoff totals at a number of sites. These were found to correspond reasonably well in the case of the Guinea stations (Fig. 5.6, for example), generating curves with a slope of about 45° and an intercept of about 200–300 mm. This corresponded well with the similar situation in the Betwa basin. In the case of the stations in Togo and Benin, the correlation between net rainfall and runoff was reasonable, but the slope was less than 45°. This response was attributed to the fact that much of the area has two rainfall seasons divided by a drier period, so that the soil moisture recharge is not constant from year to year.

Isohyetal maps of gross and net rainfall were drawn (Fig. 5.7(a) and (b)), and mean annual net rainfall was compared with mean basin runoff depth for those sites with reasonably long records (Fig. 8.23). These showed that mean runoff corresponded with mean net rainfall, less 300 mm soil moisture recharge, in Guinea. Annual runoff series could be estimated from seasonal net rainfall series with an allowance of 300 mm for annual soil moisture recharge.

In a later study of the water balance of Sierra Leone, where the climate and the seasonal distribution of rainfall were similar to Guinea, it was found possible to make reasonable estimates of average annual runoff by comparing average rainfall given by an isohyetal map with estimates of potential transpiration. The runoff estimates prepared using the methods developed for Guinea were consistent with the sparse measurements of river flow in Sierra Leone.

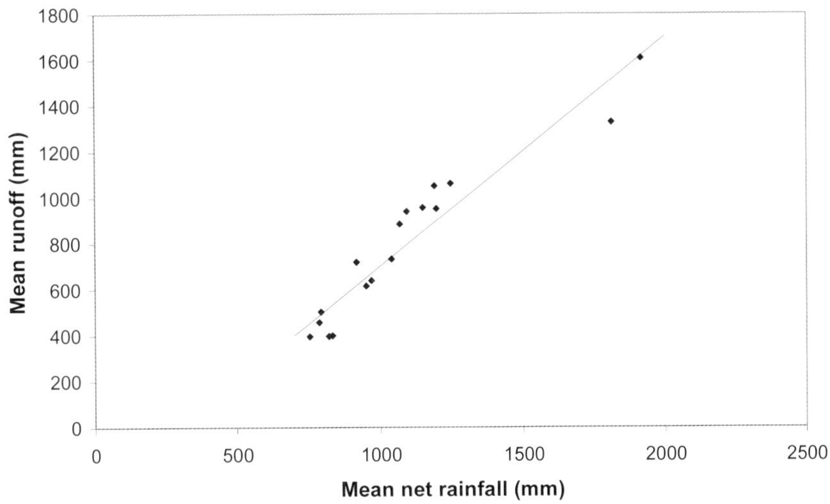

Fig. 8.23 *Guinea: mean annual net rainfall and runoff (after Sutcliffe & Piper, 1986).*

Summary

The simplifying assumption of a constant annual soil moisture recharge can only be applied where the monsoon rainfall can be treated as a single season without a significant period of drought; this is in accordance with the physics of the water balance. These examples of basins in different climates have shown that water balance studies are comparatively simple and most useful in humid areas and in areas of single-season monsoon climates, but are more difficult to apply in semiarid areas.

SOME PRINCIPLES OF HYDROLOGICAL MODELLING

Some examples of hydrological models have been introduced as part of the discussion of water balance assessment. These examples have all tended to simplify the processes involved, and it can be argued that this is a great advantage when dealing with developing countries where data are usually confined to rainfall records, meteorological data from which potential transpiration or evaporation rates may be estimated, and flow records which may not always be numerous or of long duration. Some advantages of simple models and problems of parameter estimation are discussed in this section.

Simple conceptual models

In an early example, Nash & Sutcliffe (1969) argued for simple conceptual methods in forecasting as a compromise between the assumption that streamflow could be separated into baseflow and storm runoff, deriving from effective rainfall, and the argument that it is sufficient to apply physical laws to the physical data of a given catchment in order to forecast discharge. It was noted that river flow could be forecast from rainfall and the present state of the basin, but that practical needs favoured a model based on previous history, in which the present state is itself forecast from the history of rainfall, potential evaporation and observed or computed discharge.

It was argued that if a model is to help understanding of the working of the hydrological cycle and the relative importance of the different elements, it is essential to obtain a guide to the relative significance of model parts and parametric values. It is desirable that a model should have as few parameters as possible, and in particular that two or more parts with similar effects should not occur in the same model; on the other hand, each additional part of a model should substantially extend the range of application of the model. If certain functions of the model could be isolated and its parameters optimised separately, a higher accuracy of determination should be obtained.

Ray at Grendon Underwood

This approach was applied to the records of the Ray at Grendon Underwood, England, where an experimental basin had been established in 1962 on a small area (19 km^2) of deep clay in the upper Thames to allow the development of forecasting techniques.

Occurrences of substantial rainfall were noted over a four-year period, and were related to the runoff volumes estimated between successive recession curves. A simple regression of runoff in 221 periods on the corresponding rainfall was obtained to judge the adequacy of subsequent models; the slope of the regression was found to be 0.31 with a standard error of 0.026.

The first model to be tested assumed a vertical stack of horizontal soil layers, each containing 25.4 mm of moisture at field capacity, with a total storage *Z*. Evaporation from the top layer occurs at the potential rate, then from the second layer at the potential rate multiplied by *C*, and so on. When rainfall exceeds evaporation, a fraction *H* of the excess contributes to generated runoff, and of the remainder anything in excess of a threshold value *Y* also contributes to generated runoff; the remainder percolates down to soil moisture storage. The potential evaporation is estimated from the Penman formula and multiplied by a parameter *T* to allow for systematic error. Thus the model has five parameters, some of which could be made inoperative.

After further discussion of the fitting of a model and a proposal for the definition of the efficiency of a model (Nash & Sutcliffe, 1970), this model was then compared (Mandeville *et al.*, 1970) with a second soil moisture model based on a study by Penman (1950) of the Stour; this study has been described in Chapter 6. This model divided the basin into three areas of grass, woodland and a riparian zone, and assumed that in each zone vegetation transpired at the potential rate until the wilting point was reached throughout its rooting depth and thereafter transpired at one tenth of the potential rate. The freely available soil moisture depended on the depth of rooting and soil characteristics or "root constant", and was allowed to vary between the areas of grass remote from streams, areas of woodland with deeper rooting range, and the riparian zone where available water was assumed unlimited. The total area of the basin (18.56 km^2) and the area of woodland (3.44 km^2) were known. The area of grass remote from the river (*A*) and its root constant (*D*) provided two parameters. The root constant of the woodland was found to be irrelevant in this climate. As in the other model, the ratio (*T*) of potential evaporation to open water evaporation was usually fixed. The elements corresponding to *H* and *Y* were at times used in this model also. The basis of this model is illustrated in Fig. 8.24.

It had been hoped, perhaps naively, that comparisons of these two soil moisture models would throw some light on which model more nearly represented reality. However, they were in fact similar in their operation. Both allowed the ratio of actual to potential evaporation to fall as the soil moisture deficit increases. However, the

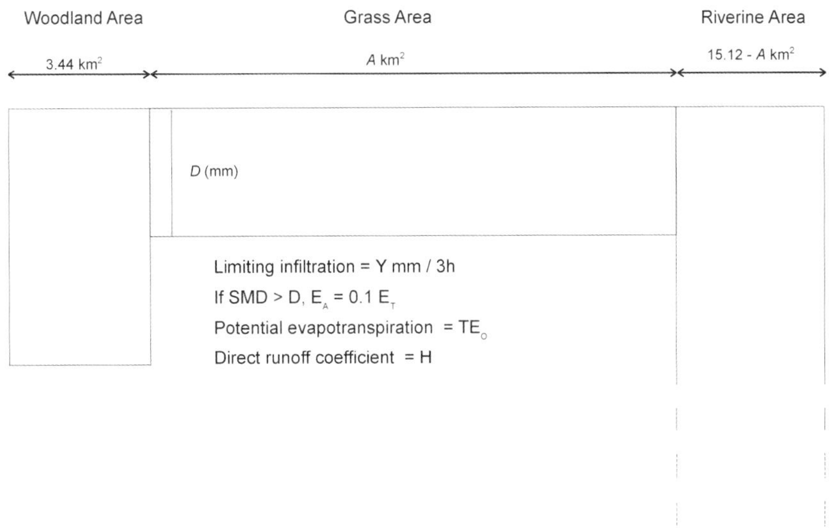

Fig. 8.24 *Basis of the conceptual model.*

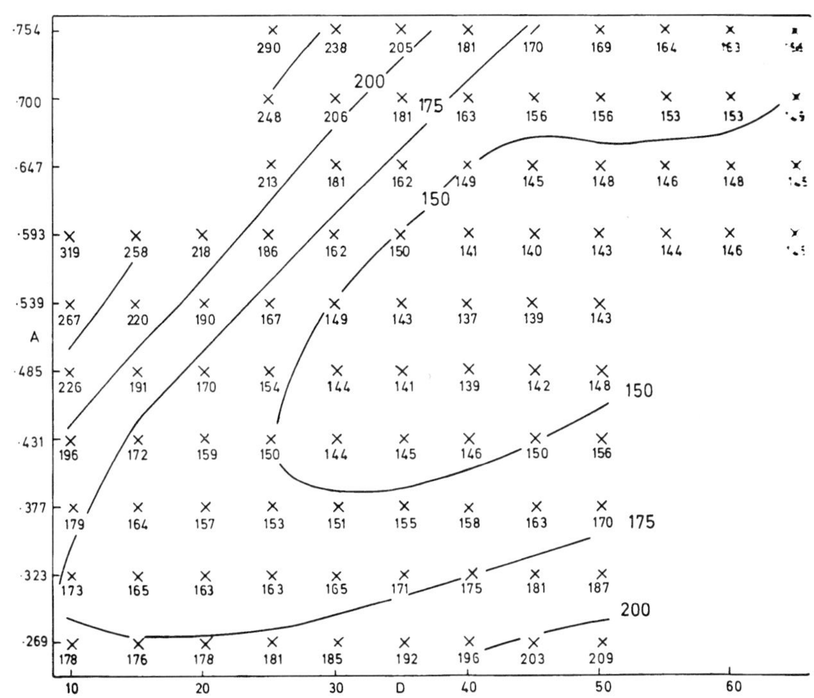

Fig. 8.25 *Dependence of the error function on A and D (from Mandeville et al., 1970).*

second model seemed to be more flexible as it allowed evaporation almost to cease from the grass area while the deficit continued to build up elsewhere.

The regression of the runoff amounts on the corresponding rainfall depths accounted for 59% of the original variance. The same data were used in a trial of the standard forecasting tool of the period, a coaxial graphical correlation relating runoff to rainfall depth, calendar month and antecedent precipitation index, which explained 81% of the initial variance. A series of trials were carried out with the two models, and for each trial the starting and optimised values for each parameter, the optimised variance, the efficiency and the total computed runoff were listed. It was found that with the second model, with one parameter fixed, and with three optimised, plus the initial deficit, the efficiency reached over 90%, which was a distinct improvement over earlier techniques.

However, comparisons of the error functions corresponding with different values of two parameters, with the other parameters fixed at reasonable values or made inoperative, showed that there was clear interdependence between pairs of parameters. In some cases the results of the optimisation, which was basically that developed by Rosenbrock (1960), were affected by starting values and local minima were reached. Figure 8.25 shows the error function surface for the second model as a function of A, the grass area, and D, the root constant. There is a valley with a flat floor, with no well-defined optimum, running diagonally across the plane. In such cases, it is possible to fix one parameter at a reasonable value after inspection of the basin, though it may be misleading to assume that the parameters correspond to their physical counterparts.

The conclusions of this comparative study were that it is possible to predict river flows with reasonable precision from rainfall and evaporation records provided that river flow data can be used to calibrate a simple model; that the calibration of a model is made easier if the model is designed to exclude parameters with similar effects on forecast flows; that proliferation of parameters requires additional measurements if their calibration is to be realistic; that it may not be possible to choose between different interpretations of hydrological processes by comparisons of models alone.

WATER BALANCE OF LAKES

Studies of the water balances of lakes have advantages and problems of their own. Because lake levels are often available for periods before other hydrological records were started, the study of lake water balances can extend the period of information about the water resources of a region. Where lakes form the limit of an internal drainage system, lake levels can give direct indicators of variations of water resources before the availability of inflow and outflow records. This has been true of Lake Turkana, which receives inflow from the Omo valley in Ethiopia and from the Turkana River in northern Kenya. Historic levels have been inferred from investigations by Butzer (1971) who was able to use the maps compiled by early travellers. These levels support similar trends in other parts of East Africa. The former levels of other lakes in the Ethiopian Rift Valley have been investigated by Grove *et al.* (1975) and given insight to climatic change over this key area. Even where lakes have outlets which have only been recorded for limited periods, they can provide a useful check on regional hydrological records. During a study in 1986 of the water resources of rivers flowing into Lake Abaya in the Rift Valley, it was possible to use lake levels to confirm the validity of flow records and also the derivation of rainfall–runoff relations used to infer the flows of minor tributaries. Nearly complete records of inflow to the lake were compared with lake rainfall and evaporation to provide predicted lake levels over a period of nine years; when these were compared with measured lake levels a period of apparent inconsistency was noted, but this was found to coincide with a rise in lake level which resulted in a period of high flows recorded in a river flowing from Lake

Abaya to Lake Chania downstream. Thus the downstream lake evidence supported the validity of the upstream water resources study.

There are channels leading from these two lakes towards Lake Stefanie, which is normally an area of grass and wetland; there is evidence that on rare occasions Lake Stefanie also becomes a true lake, and these periods coincide with high levels of Lake Turkana and Lake Victoria, which are known to have occurred around 1878 and 1961.

Lake Victoria

The water balance of Lake Victoria has been the subject of a number of studies because of its importance as the source of the White Nile, and because of the dramatic rise in lake level and outflow which occurred after 1961–1964.

Lake levels have been measured regularly since 1896. From 1896 to 1960, the levels fluctuated between 10.2 and 12.0 m on the Jinja gauge (Fig. 8.26), but rose 2.5 m to over 13.0 m in 1961–1964. Since 1964 the general trend has been downwards, but there were rises in 1979 and 1998. The lake outflows were controlled naturally by the geometry of the Ripon Falls before 1951, and the levels and outflows were linked by a stable control. Since 1951, when the construction of the Owen Falls dam began some 3 km downstream of the Ripon Falls, the outflow has been controlled to match the earlier relation between lake level and outflow, so that the dam is operated on a run-of-river basis. This relation was deduced from current meter gauging below the lake, but was extended by hydraulic modelling of site conditions to match higher levels. A composite outflow record has been deduced (Sene & Plinston, 1994) and is brought up to date in Fig. 8.27.

Some 20 tributaries into Lake Victoria drain a total area of 190 000 km^2 compared with a lake area of 67 000 km^2. The Kagera, draining much of Rwanda and Burundi, has been measured since 1940. Several Kenyan tributaries have been measured since 1956; nearly all of the other tributaries were measured from 1969 to 1978. The total tributary inflow for 1956–1978 was estimated from these records. The inflow (Fig. 8.28)

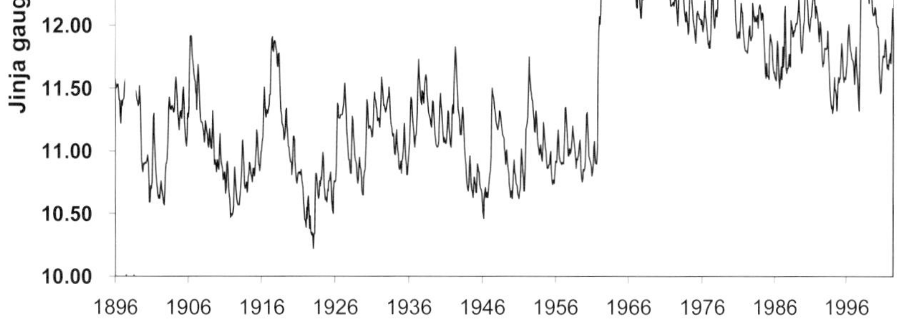

Fig. 8.26 *Lake Victoria lake levels: Jinja gauge, 1896–2000.*

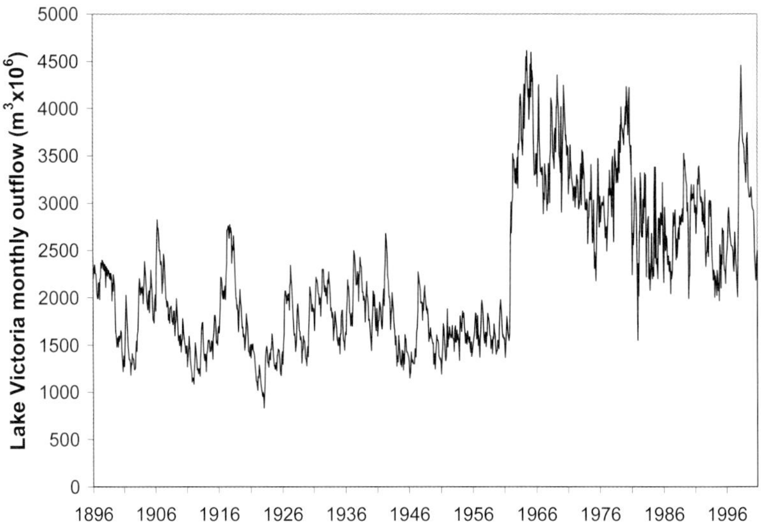

Fig. 8.27 *Lake Victoria outflow: 1896–2000.*

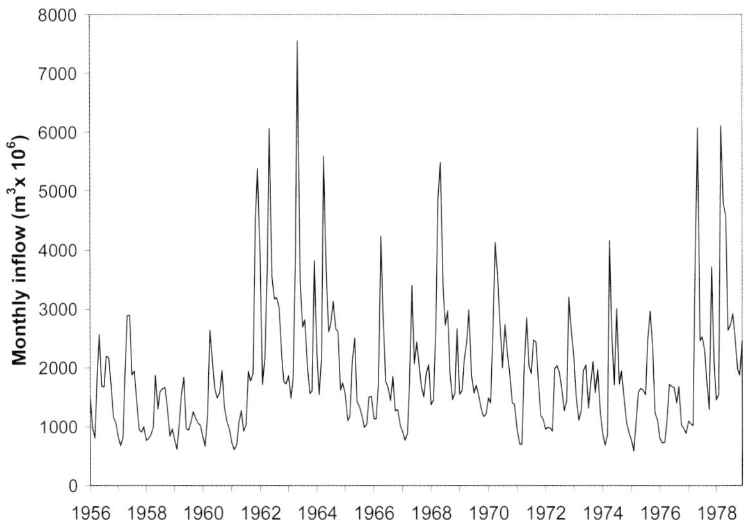

Fig. 8.28 *Lake Victoria tributary inflow: 1956–1978.*

increased markedly after the heavy inflows of 1961–1964. This inflow was variable from year to year as the natural variability of catchment rainfall was supplemented by the sensitivity of the runoff coefficient to rainfall.

The main component of the input to the lake is in fact the rainfall over the lake surface; indeed this rainfall has been described as the true source of the Nile. The interaction of lake breezes with the prevailing easterlies results in higher rainfall over the lake than on the surrounding basin, with rainfall in the morning on the eastern shore and evening rainfall on the west. This has been confirmed by some measurements on islands in the lake. It is not easy to develop a long-term series of lake rainfall, as the only continuous stations are at eight sites around the lake.

However, a reasonable series was deduced (Piper *et al.*, 1986) by using these stations as indices of monthly and annual rainfall, while enhancing the variability to counteract the effect of averaging the indices, and multiplying the series by the mean lake rainfall derived from the water balance. There is evidence from the rainfall records around the lake that there was a significant rise of rainfall after 1961–1964 in the October–November season, which has evidently been sufficient to affect the lake balance.

As it is difficult to distinguish variations in rainfall from those of lake evaporation, the lake evaporation was taken as constant from year to year, and was based on monthly average open water evaporation derived by the Penman method for stations around the lake.

The annual means of all these water balance components for the period 1956–1978 are expressed in Table 8.2 as mm over the lake area, and their variations are expressed in terms of standard deviation and coefficient of variation. Table 8.2 confirms that rainfall is the major component of the balance, but also that tributary inflow, though only about 15% of the total input, has a much higher relative variability and is therefore an important factor in the rise of the lake. The lake balance is extremely sensitive to increases of lake rainfall and related tributary inflow, as the average rainfall is of the same order as lake evaporation; an increase in input has a disproportionate effect on the outflow, which has a variability somewhat higher than the inflow alone.

Table 8.2 *Annual water balance of Lake Victoria, 1956–1978 (after Sutcliffe & Parks, 1999).*

Year	Lake rainfall (mm)	Tributary inflow (mm over lake)	Lake outflow (mm over lake)
Mean	1858	343	524
SD	180	103	149
CV(%)	9.7	30.1	28.4

Table 8.3 *Monthly components of Lake Victoria water balance, 1956–1978 (after Sutcliffe & Parks, 1999).*

Jan	Feb	Mar	Apr	May	Jun	Jul	Aug	Sep	Oct	Nov	Dec	Year
Lake rainfall (mm)												
138	153	228	309	208	71	52	68	104	157	201	170	1858
Total inflow (mm over lake)												
20	16	23	41	48	31	30	31	29	23	25	26	343
Lake evaporation (mm)												
135	135	145	130	125	120	125	135	140	145	130	130	1595
Lake outflow (mm over lake)												
43	39	43	44	47	47	46	44	43	43	41	44	524
Lake level (m on Jinja gauge)												
11.93	11.93	11.99	12.15	12.23	12.16	12.06	11.98	11.92	11.90	11.97	12.01	

The seasonal lake balance is summarised by Table 8.3, where the components of the balance and the resulting lake level and outflow are given. The annual balance is illustrated by Fig. 8.29, where the observed and predicted annual lake levels are compared. It will be seen that the balance is not only close over the whole period, which is inevitable from the method of deriving rainfall, but also follows the pattern of rises and falls between these dates. The result of the change in the lake balance has been the doubling of the lake outflows after 1961–1964. This rise has been reinforced by the net effect of the basins of Lake Kyoga and Lake Albert (Sutcliffe & Parks, 1999), and has therefore had a dramatic effect on the wetlands of the Sudd downstream.

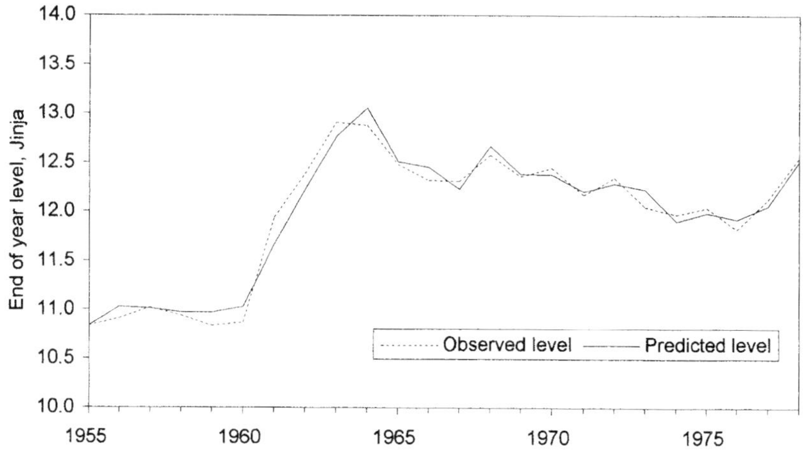

Fig. 8.29 *Lake Victoria annual balance, 1956–1978 (from Sutcliffe & Parks, 1999).*

It has been possible to use indirect evidence to take the study of Lake Victoria hydrology back before the start of historic lake levels. Lyons (1906) had visited the East African lakes and gathered oral and published evidence at that time about earlier lake levels. There were a number of travellers' observations of exceptionally high lake level in August 1878, a fall of about 2.5 m by 1891, followed by a rise in 1892–1895 and a period of falling levels from the start of observed levels in 1896. The 1878 peak was supported by observations of flooding at Lado below Lake Albert, where levels were still high in December 1878.

These high levels were supported by evidence from downstream. There is evidence of widespread flooding in the Sudd in 1878 and 1895, while river levels at Khartoum in May 1879 indicated high lake outflows in 1878. Observed flows at Aswan were high on several occasions in the period 1870–1898, and it has been possible to reconstruct early levels from downstream evidence (Tate *et al.*, 2001). There is archaeological evidence for an upper bound to lake levels (Bishop, 1969). The 1964 peak levels were within 60 cm of the level of gravel deposits with dated charcoal fragments in Hippo Bay cave. Bishop (1969) concluded that "at no time since 3720 ± years ago has Lake Victoria risen to re-occupy the 10- to 12-foot cliff notch" (a level equivalent to 13.85 to 14.46 m on the Jinja gauge).

It is also possible to link modelling of Lake Victoria to climate changes in rainfall and evaporation to estimate potential impacts on lake water balances and levels (Tate *et al.*, 2004).

WETLAND WATER BALANCE

Wetlands play an important role in a number of major river systems. They have received considerable attention in the Nile basin, where the loss to evaporation of about half the inflow of the Bahr el Jebel before it reaches the White Nile has drawn attention to the hydrology of the Sudd. Proposals to divert the part of the inflow to the Sudd around the swamps in the Jonglei Canal have been studied in various forms since 1904, and the benefits and the impacts of this project are considered in Chapter 11. Here, the problems of carrying out water balance studies in major wetlands are discussed and illustrated by studies of the Sudd and other major African wetlands.

The major difference between water balance studies of wetlands and lakes are that the transpiration from swamp vegetation has been less studied than open water evaporation, and therefore there is less certainty about the rate of transpiration or its relation to open water evaporation. Methods are now available to measure evaporation rates above different surfaces, and these should in the future provide means of refining wetland water balances; however, up to the present it has been necessary to apply the general principles of water balance. It is also true that while the relation between lake area and volume of water stored can be deduced from surveys around the lake, together with depth surveys within the lake, it is not a simple matter to derive relations between wetland area and water volume, even with sophisticated methods of survey. The problem is similar to deriving area–volume relations for a reservoir, with the added problems that access is not easy, that water surfaces can vary in level over an area, and that both water surfaces and water depth are usually obscured by vegetation. Indeed, the water depth in a papyrus swamp is not well defined, and even when using a probe must be judged by a change of resistance. It is likely that a comparison of swamp area, revealed by satellite imagery, with volumes stored derived from water balance studies, could now present a means of deducing area–volume relations, but it may not be easy to find periods when satellite availability and precise swamp water balances coincide.

The Sudd

In the case of the Sudd, there is some evidence from surveys that a linear relation between area and storage is reasonable. This is equivalent to an exponential relation between water surface elevation and area, and thus to a constant average depth of water over the flooded area. This is not an unreasonable simplification; it has the incidental advantage that there is in this case a direct solution (Sutcliffe & Parks, 1987, 1989) to the water balance equation linking successive areas and volumes of flooding, whereas other forms of relationship require an iterative solution to allow for evaporation from a changing area.

The equation of continuity for a time interval dt is:

$$\mathrm{d}V = [Q - q + A(R - E)]\mathrm{d}t - r\mathrm{d}A$$

where V is volume of flooding, Q is inflow, q is outflow, R is rainfall, E is evaporation, A is flooded area and r is soil moisture recharge, which is positive but set to zero when dA is negative. Percolation to groundwater is unlikely to be significant in the case of the Sudd. Field evidence suggests that a linear relation between V and A is reasonable for the Sudd, and this can be expressed as $A = kV$ as $A = 0$ when $V = 0$. This implies a constant average flooding depth of $1/k$ m, and a value of $k = 1$ was deduced from personal observation and adopted after testing different values and also different forms of the relationship.

The analysis for month i, allowing for rainfall and evaporation over the mean of the initial and final values of area A, and substituting kV for A, becomes:

$$V_{i+1} = V_i + Q_i - q_i - k/2(V_i + V_{i+1})(E_i - R_i) - kr_i(V_{i+1} - V_i)$$

and hence

$$V_{i+1} = (V_i[1 + k\{r_i - (E_i - R_i)/2\}] + Q_i - q_i)/[1 + k\{r_i + (E_i - R_i)/2]$$

In the case of the Sudd, monthly inflows and outflows were known from 1905 to 1980, and a rainfall series was compiled from nine stations around the swamps. Evaporation

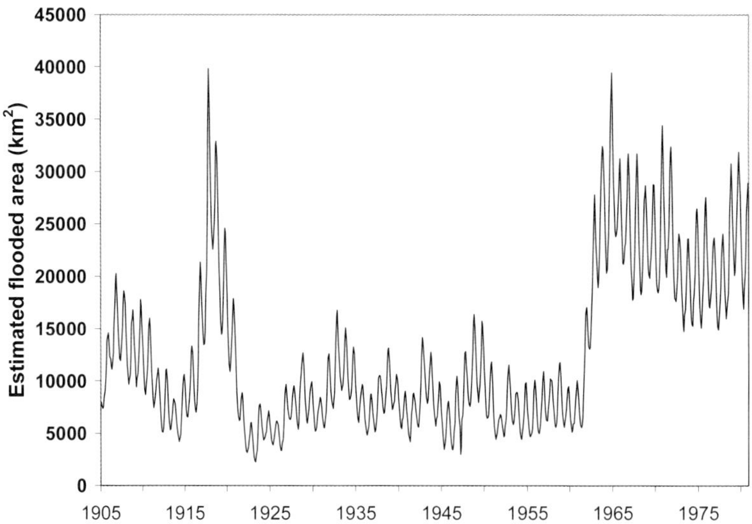

Fig. 8.30 *Bahr el Jebel swamps: monthly estimates of flooded area, 1905–1980 (after Sutcliffe & Parks, 1987).*

was taken as open water evaporation estimated for Bor by the Penman method. Soil moisture recharge was taken as 200 mm at the start of the wet season, and reduced by net rainfall during the wet season.

Storage volumes and flooded areas were estimated at monthly intervals starting from initial estimates of 8000 $m^3 \times 10^6$ and 8000 km^2, respectively. These values corresponded reasonably with estimates of flooded areas derived over the years from survey maps, vegetation maps and satellite imagery. The series of monthly flooded areas (Fig. 8.30) showed the influence of the seasonal torrents in providing annual areas of flooding and recession, which provide the important areas of seasonal flooding, while the lake-fed steady flows of the Bahr el Jebel below Lake Albert are responsible for the more permanent swamp. However, the important effect of the rise in Lake Victoria after 1961–1964, when the areas of permanent swamp more than doubled, is clearly evident.

African wetlands

A similar analysis was applied (Sutcliffe & Parks, 1989) to provide a comparison between four major African wetlands: the Sudd of the Bahr el Jebel; the Okavango swamp in Botswana; the Niger interior delta in Mali; and the wetlands of the lower Senegal basin. The main differences between the four wetlands were the seasonal distributions of river inflow. Those of the Sudd were a combination of lake-fed steady flows and highly seasonal torrents. The inflow to the Okavango was seasonal but damped in the long course from the mountains of Angola. The inflows of the Niger and Senegal are both derived from the highly seasonal rainfall in their headwaters in the mountains of the Fouta Djallon in Guinea; they are therefore markedly seasonal with the flows concentrated in 3–4 months.

The inflows and outflows, together with average rainfall and estimates of open water evaporation, were used in the model developed for the Sudd to generate

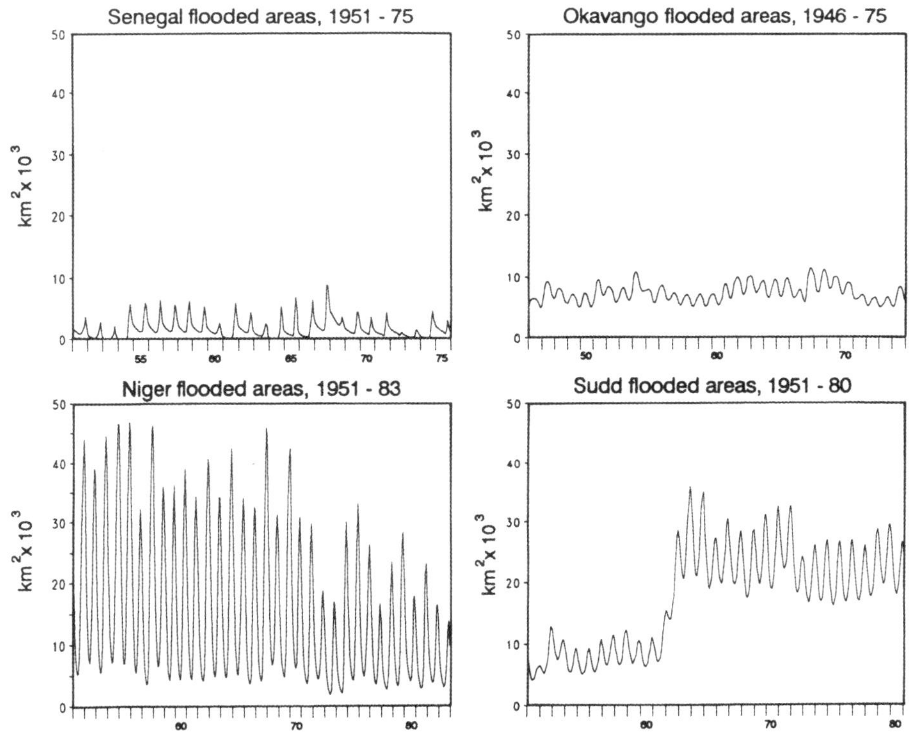

Fig. 8.31 Comparison of flooded areas (km² x 10³) in Senegal, Okavango, Niger, Sudd (from Sutcliffe & Parks, 1989).

estimates of flooded areas over similar numbers of years. The only difference was that an average depth of flooding of 2 m rather than 1 m was assumed for the Senegal wetland. The results (Fig. 8.31) showed that what appeared to be areas of similar status on the map had quite different flooding regimes. The Okavango swamp had similarities with the Sudd, in that there was a significant area of permanent swamp, though the effect of Lake Victoria was missing. The wetlands of the Niger and Senegal, however, were much more seasonal with small areas of permanent flooding. In contrast to the increased flooding in the Sudd after the rise in Lake Victoria, the Niger and Senegal wetlands have decreased in extent after about 1970, as a result of the effects of the Sahel drought on river inflows.

The vegetation corresponding to wetlands of different hydrological regimes is discussed in Chapter 11; however, the contrast between the flooding regimes of these major African wetlands is reflected in their vegetation. Those with high proportions of permanent swamp are naturally more dominated by papyrus, while those with a higher proportions of seasonally flooded land are dominated by grasses. These different regimes are controlled by the inflows to the wetlands rather than the local climates. The Sudd and Niger wetlands are similar in terms of latitude and seasonal climate; the key difference is the large lake-fed inflow to the Sudd which feeds the permanent swamp.

CONCLUSION

These examples illustrate the general point that simple water balance models can be very effective in explaining the physical processes operating in a river basin, and can lead to an understanding of the runoff provided that the climate, the physical hydrology and the vegetation are carefully considered. Comparisons of precipitation and potential transpiration can lead to a simple understanding of different hydrological regimes, but the seasonal timing is as important as a comparison of annual totals. Provided that an understanding of the main driving forces is incorporated, simple models of the water balance can be extremely useful in explaining the hydrology of an area, and can provide a useful check on more complex hydrological modelling and on the basic data. In fact in these days of increasing complexity of analysis and increasing reliance on computer manipulation of data, it is extremely important to ensure that the hydrological forces controlling runoff are understood and incorporated in analysis, rather than relying on more complicated sets of parameters and relationships in attempting to explain runoff.

9 SURFACE WATER RESOURCE ASSESSMENT

The assessment of water resources for water supply, irrigation, or hydroelectric power is one of the main engineering applications of hydrological studies. The water resource development may be based on direct abstraction or may require reservoir provision and storage. The assessment may be limited to a single site or may involve the comparison of a number of sites, which will depend in the first instance on estimates of the mean annual river flow. The following examples of such assessments are designed to illustrate some of the variations encountered.

The importance of using as long a period as possible to estimate the mean river flow leads to the practice of extending flow records by correlation with longer flow records, or by comparison with rainfall records. Regression may be based on actual records, or may be preceded by logarithmic or other transformation. Examples of record extension are drawn from various climates. The analysis of yield from river sites can vary from simple run-of-river analysis to the assessment of yield from a storage reservoir. Reservoir analysis can vary from historical trial of the storage required to maintain a given yield, to statistical derivation of the relation between reservoir volume and yield with a given frequency of success. Examples of various approaches used in Botswana are described.

ASSESSMENT OF MEAN RUNOFF

The average river flow is the most commonly used measure of comparison between sites, and is a key element in the studies of water balance which take account of all the hydrological processes within a basin. It is possible to use flow measurements alone to estimate this quantity, but the reliability of the estimate is improved if it is tested by the assessment of a complete water balance. Although the mean annual river flow is a simple basis of comparison, its estimation should include both seasonal variation and long-term variability. It is essential to ensure an average which is consistent between stations, and this implies an average over as long a period as possible covering a common series of years.

The period of flows should sample the variety of droughts to which the region is liable; in Africa in particular the average flow has been shown to vary significantly from one period to another. The Sahel drought which started in the 1970s has demonstrated that river flows, even when based on periods of record of up to 30 years, can be seriously misleading as criteria for design. Hydroelectric schemes in Côte d'Ivoire, for example, could be designed in the 1960s on the basis of records which should have been adequate to sample the runoff regime but have in practice overestimated the potential yield in subsequent periods. In some cases, river flows for the period 1971–1985 were only 50% of the average for the preceding 40 years.

Similar changes have occurred in both directions in the Nile basin. The behaviour of the White and Blue Niles has provided contrasting features which illustrate the importance of basing estimates on long-term records. The White Nile flows were relatively steady from 1896, when Lake Victoria level records and deduced outflows began, until 1961. However, between 1961 and 1964, the lake rose much higher than in the previous period and the outflows doubled. Since that event, the lake levels have remained relatively high, though the trend appears to be downwards. Subsequent study of early accounts of the lake level has revealed that high levels and outflows also occurred in the last quarter of the 19th century. The Blue Nile, by contrast, has suffered droughts in recent years which have been similar to those of the West African rivers affected by the Sahel drought. At the lower limit of the Nile, where flow volumes are dominated by the Blue Nile, the flow records at Aswan and other sites reveal periods of high flows from 1870 to 1900, of moderate flows from 1900 to 1970, and of low flows since 1970. The flows measured during the season when White Nile flows are dominant, however, confirm the high lake levels and outflows in 1878–1895 and after 1964. It is clear that longer periods of record are essential in these climates than had been thought necessary in Europe where 30-year periods of rainfall record were used to illustrate long-term regimes.

In the light of recent awareness of climate change it is even more essential that the longest flow record is used in order to reduce the errors of estimation of the mean and variability. One method of maximising the effective period of record is to extend the river flow record by correlation.

EXTENSION OF RIVER FLOW RECORDS

It is inevitable that in the establishment of hydrological networks, some types of record and some stations will be established before others. In general precipitation records cover a longer period than river level or flow records, as a raingauge is simpler to install and maintain than a flow record. There are exceptions, as some river level records were used for navigation and could later be adapted for hydrological purposes; there are examples on the River Karun in Iran and on the River Senegal.

It is also usual that river gauges are established on major rivers and near major cities, before they are installed on smaller streams. There is therefore usually scope for extending flow records by correlation with longer established stations in the same basin, where flows should be reasonably closely linked. There are also opportunities for extending flow records by correlation with precipitation records, which often cover a reasonable period of time.

Correlation techniques are described in many textbooks; for example applications in hydrology are discussed in Holder (1985). One characteristic of hydrological data is that they are commonly skewed, with a few large values contrasting with many small values. It is often desirable to transform the data, for instance to logarithms of the actual data, before undertaking regression analysis.

Extension of river flows in the Senegal basin

In a study of the control of the River Senegal by the OMVS (Organisation pour la Mise en Valeur du Fleuve Senegal), following the construction of a dam at Manantali in the headwaters, it was necessary to compile a set of records at key stations on the river network (Figs 3.1 and 9.1) providing a consistent series covering the longest possible period. After visits to the key hydrological offices at Dakar and Saint Louis in Senegal, at Bamako in Mali, and at Nouagchott in Mauritania, and to the ORSTOM offices in Paris and Dakar, the key records at three stations in Senegal and 11 in Mali were available for analysis. The flows at the key station of the Senegal at Bakel were corrected by ORSTOM to take account of hysteresis in the rating curve; because the water level gradient is steeper on a rising than a falling river, the rating curve is looped.

There were two main objectives in the analysis (Sutcliffe & Lazenby, 1989). The first was to extend the inflow record for the site above the reservoir at Manantali in order to simulate the operation of the reservoir over a representative period. The other was to derive a consistent set of flows of the various tributaries in order to study the contributions of each to the downstream flow. To meet the first objective, it was necessary to fill gaps in the flow record at Soukoutali, just below the reservoir site, and to extend the flows back to 1904 using the records at Kayes and Bakel downstream. For the second, the objective was met by completing a set of records spanning the period 1952–1984, as a number of stations covered this period. For the first exercise there was a choice of only two stations for extension, but the second exercise could draw on a wider selection.

It is common practice to introduce a stochastic element into the regression, as in the case of the Blue Nile tributaries discussed below, in order to preserve the variance of the measured flow record. In this case, as the stations to be completed were upstream of Bakel, the long-term station, the introduction of random errors would introduce a lack of correspondence between the upstream and downstream stations, and so it was preferred to use a single regression equation. A linear equation for each calendar month was used to extend or complete each station. This was necessary because the proportion of river flow coming from each tributary varied between the months of both high and low flow seasons. The station to be used for extension was selected by correlating each set of monthly flows for each pair of stations to give correlation coefficient (r) values, adding together the 12 r values for each pair of stations, and using the pair of stations with the highest sum. In general the station selected in this way for extension proved to be the nearest station downstream or upstream, but there were exceptions, including a case where the choice for the independent station was the difference of flows between two stations.

The mean annual flows for the different tributaries were combined in a diagrammatic representation of the basin behaviour (Fig. 9.1) and the long-term mean flows of the tributaries were compared with mean basin rainfall in Fig. 9.2 to illustrate the water balance regime of the area. The scale of the variations of annual flows at Bakel is shown in Fig. 9.3, where the effect of the drought of recent years is evident. It was interesting to note that the variations in river flows from year to year were, as expected, greater on the tributaries draining drier areas (see Fig. 3.1). Thus the proportion of the main river flow passing through the reservoir at Manantali on the Bafing, which was the wettest basin, is higher in dry years. Conversely, the proportion of flow on this tributary is lower in years of high floods. This implies that the reservoir is well placed to store water in years of low flow, but is not so capable of controlling floods in years of high flows.

In a later study (Lazenby & Sutcliffe, 1994), these flows were used to investigate how the Manantali reservoir could be operated to control the total flows of the River

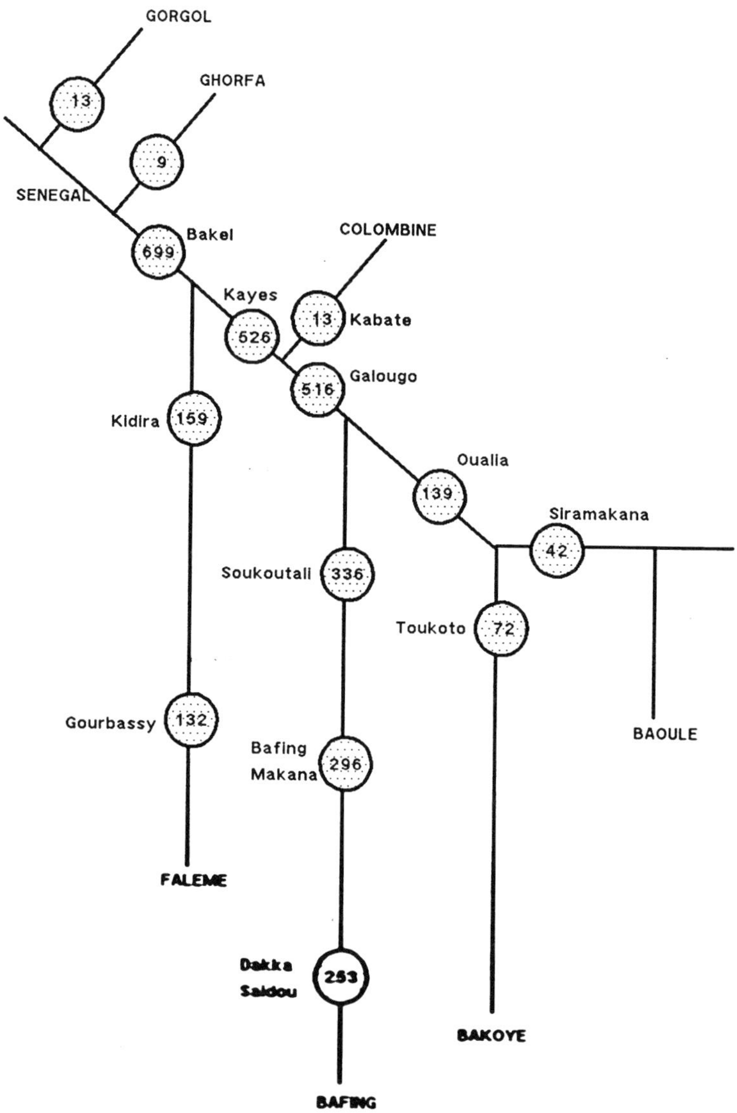

***Fig. 9.1** Senegal basin: representation of mean annual flow 1952–1984* ($m^3\ s^{-1}$).

Senegal. As an interim measure it was proposed to release water from the reservoir to simulate the natural inundation of the flood plain in the lower Senegal. This was used for grazing and natural irrigation or recession culture, exploiting the soil moisture recharged by the flood. The design of this artificial flood, in terms of its shape and duration, took account of traditional practice, and its combination with the hydro-electric, irrigation and navigation components of the project was studied using an operational model of the reservoir.

It would clearly be necessary to monitor flows on the other tributaries in order to calculate the releases from Manantali which would give the desired combined flow hydrograph downstream. Preliminary tests of the reservoir control were based on

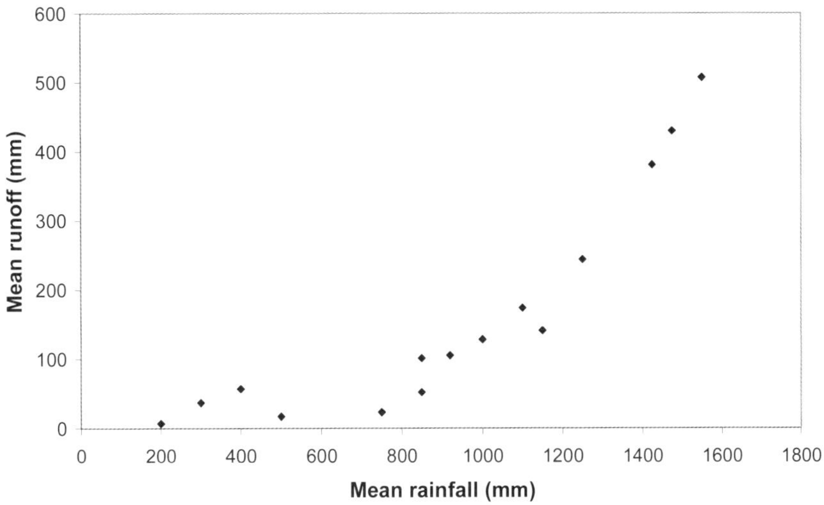

Fig. 9.2 *Senegal basin: mean rainfall and runoff depth (after Sutcliffe & Knott, 1987).*

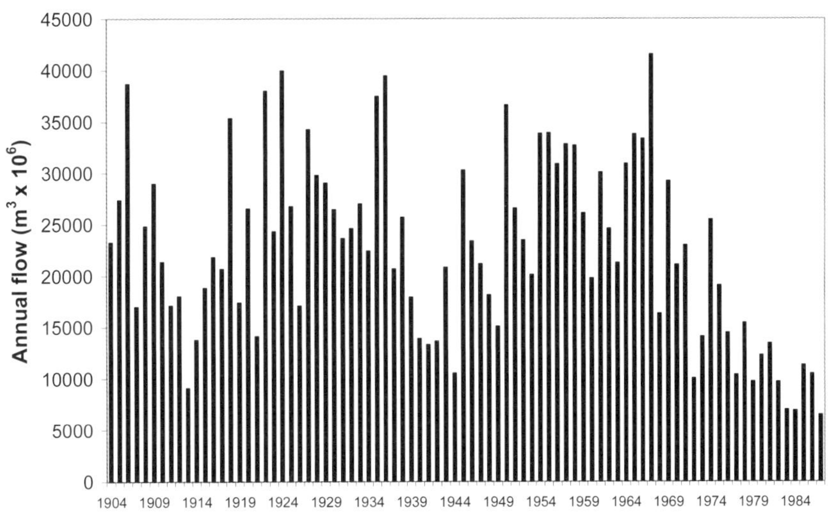

Fig. 9.3 *Senegal at Bakel: annual flows, 1904–1987.*

simultaneous daily flows at Gourbassy on the Falémé and Oualia on the Bakoyé, compared with flows at Bakel, as travel times from these stations and the reservoir were about three days. Later tests of reservoir operating rules were based on the complete record of daily flows, and the potential benefits of flow forecasting have also been studied (Bader, 1992). The actual effect of reservoir releases is illustrated in Fig. 11.12.

Extension of flows on Blue Nile tributaries

A different approach was used to extend the flow series for the Dinder and Rahad, which are relatively minor tributaries of the Blue Nile (Fig. 7.1). The three rivers drain basins in western Ethiopia, but the Blue Nile basin extends to an area of higher ground around Lake Tana where the rainfall is higher. The Blue Nile where it enters the Sudan has a small but important dry season component of flow, due to a longer rainfall season in its southern tributaries and also lake and basin storage, whereas the Dinder and Rahad are dry from January to May at the sites where they enter the Blue Nile between Sennar and Khartoum.

At the time of a water resources study (Institute of Hydrology, 1978), flows for the Dinder and Rahad were required to supplement the records of the Blue Nile, which had been measured at Roseires from 1912 and at El Deim upstream from 1962. The Dinder had been measured near its mouth from 1907 to 1951, and at an upstream site from 1972; the Rahad record was similar, with measurements near its mouth from 1908 to 1951, and upstream from 1972. The gaps in the record between 1951 and 1972 were filled by means of a statistical model, using the earlier concurrent flows measured on the Blue Nile at Roseires and the Dinder and Rahad at their mouths. To avoid including zero values, the months of low flow were grouped into seasons, and an inspection of the corresponding series led to the adoption of a logarithmic transformation before analysis. Cross-correlations between the three flow series, and lag-one serial correlations within each series, were derived from the concurrent flow records. Flow series were generated for the Dinder from the Blue Nile records, and flows for the Rahad were generated from the Blue Nile and Dinder series, preserving both the cross-correlations and the serial correlations. Because the flows were required for an analysis of joint use, and because the three rivers were independent rather than linked like the Senegal examples, a random element was included to preserve the natural variance, and a typical sequence was selected for subsequent analysis. Normally a variety of sequences might be used, but in this case the tributaries provided a minor contribution to the resource.

EXTENSION OF RIVER FLOWS USING RAINFALL RECORDS

It is common for rainfall records to predate measurements of river flows, which require a more formal organisation to maintain and calibrate river level records. As with water balance studies, the type of climate affects the ease with which rainfall data can be used. In a humid climate, the relation between rainfall and river flow is relatively simple as actual evaporation will be close to potential transpiration and will not vary much from year to year. In an arid or semiarid climate, this is no longer true, and basin evaporation is highly dependent on the incidence of rainfall; a more complex or even empirical relation between rainfall and runoff is required, and the uncertainty of the extended series is greater. Although the monsoon type climate, where there is a seasonal switch between humid and arid conditions, would appear to be complex, it is possible to simplify the situation by considering the monsoon season as a single storm. Examples of these situations are introduced to illustrate these differences.

Kenya and New Zealand

In two early investigations (Sutcliffe & Rangeley, 1960a), estimates of water resources potential for irrigation and hydroelectric power were required at sites where river flow records were comparatively short. However, rainfall records over the basins above these sites were available to extend the flow records backwards in order to estimate the mean and variability of flows over a longer time scale.

In the case of the Tana basin in central Kenya, discussed in Chapter 3, the long-term average basin rainfall was estimated as 1170 mm, but the average varied over the basin from 650 mm over the plain at the lower end of the basin to 2000 mm on the mountainous headwaters (Fig. 3.7). Some 22 local rainfall stations provided records which were divided by their individual averages to give annual indices for each station; these were combined to give basin average indices which were multiplied by the basin average rainfall to give an estimate of the annual basin rainfall for a series of 50 years. This series was compared with annual basin runoff over a common period of 11 years (1947–1957), and the annual flow series was extended to a period of 50 years by comparison. It showed (Fig. 9.4(a)) that the runoff coefficient and also the basin evaporation both increased in years of higher rainfall. It is likely that the period during which water was available for transpiration was longer during wet years; the situation is made more complex in this example because there are two separate and marked rainfall seasons. The extension made it possible to place the 11 years of flow record in a wider context; in fact it emerged that the highest and lowest years in the shorter period were also the two extremes of the longer period. During a later review of a reservoir project on the Tana, carried out in 1982, it was interesting to note that the dry year 1949 was still the critical period in the flow record; the high flows had been superseded by the higher flows after 1961.

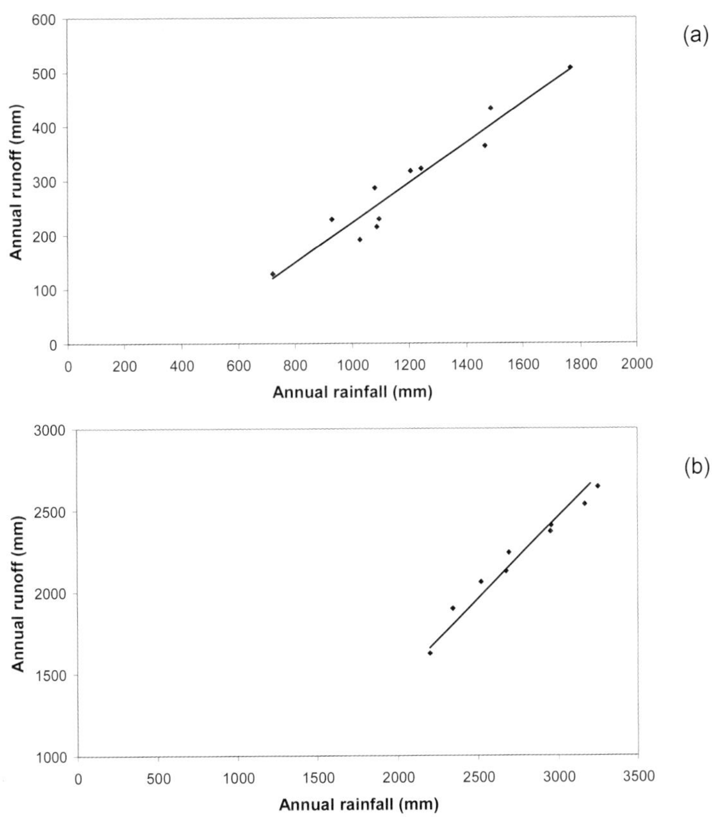

Fig. 9.4 Annual rainfall and runoff: (a) Tana, and (b) Tongariro (note the different scales of all the axes) (after Sutcliffe & Rangeley, 1960a).

The Tongariro basin, in the centre of the North Island of New Zealand, drains an area of some 780 km^2 including the volcanic massif of Mount Ruapehu (2800 m), with heavy precipitation throughout the year and falls of snow at higher elevations. The records of five stations in or near the basin showed that mean precipitation was closely linked to elevation, and ranged from about 1270 mm near Lake Taupo to 5000 mm at 1800 m; the average basin precipitation (Fig. 3.3) was estimated at about 2700 mm. Although the number of long-term stations was small, they were reasonably homogeneous, so that it was possible to derive a basin rainfall series covering 30 years, with which the flow record could be compared. The annual basin evaporation loss appeared to be fairly constant at 540 mm from year to year and the rainfall–runoff relation (Fig. 9.4(b)) was drawn on this basis. The runoff record was extended from nine years to 30 years. Because the evaporation was small by comparison with the average rainfall, the variability of the runoff was similar to that of the rainfall and was comparatively small.

Extension of West African river flows

The simple modelling of rainfall and runoff in a monsoon climate has been described in Chapter 5 with the example of the Betwa basin in central India. It was shown that the simple surplus of rainfall over potential transpiration first fills the soil moisture store and is then transformed into runoff or to a lesser extent into groundwater recharge. Thus the seasonal total net rainfall, or the sum of rainfall less potential transpiration, can be correlated with runoff; the slope should be at 45° while the intercept on the net rainfall axis should be equivalent to the soil moisture deficit at the end of the dry season. Following this study, the seasonal total net rainfall during the monsoon season was related to the annual river flows in a regional investigation (Sutcliffe & Piper, 1986) of the hydroelectric potential of Guinea, where the seasonal climate was similar. It was found that the annual series at individual stations (Fig. 5.6, for example) followed the predicted pattern, so that short river flow series could be extended by plotting the two series, and using the line of equality to estimate the runoff in periods when rainfall alone was available.

Similarly the regional relation between rainfall and runoff (Fig. 8.23) could be simply deduced by using the average value of net rainfall over gauged basins. After calculating the average net rainfall at sites with long rainfall records, a net isohyetal map was drawn from the station values and the net rainfall values for basins derived. The average river flow at ungauged sites could be estimated after subtracting the soil moisture deficit, and the annual runoff series could be deduced from basin rainfall series. As the objective of the study was to assess the potential of a number of sites with reservoir storage, the seasonal flow distribution was not critical, and the average monthly distribution at gauged sites could be applied to the annual flows.

PROBLEM OF VARIABILITY OF AFRICAN RIVER FLOWS

Another example of water resources assessment in West Africa was a study in 1986–1987 of the Bandama basin in Côte d'Ivoire. After visits to the key gauging stations to review the record quality, and collation of flows, the water balance was assessed of tributaries flowing from the north towards the south of the country. In order to derive consistent series, the river flows were extended to a common period by correlation with

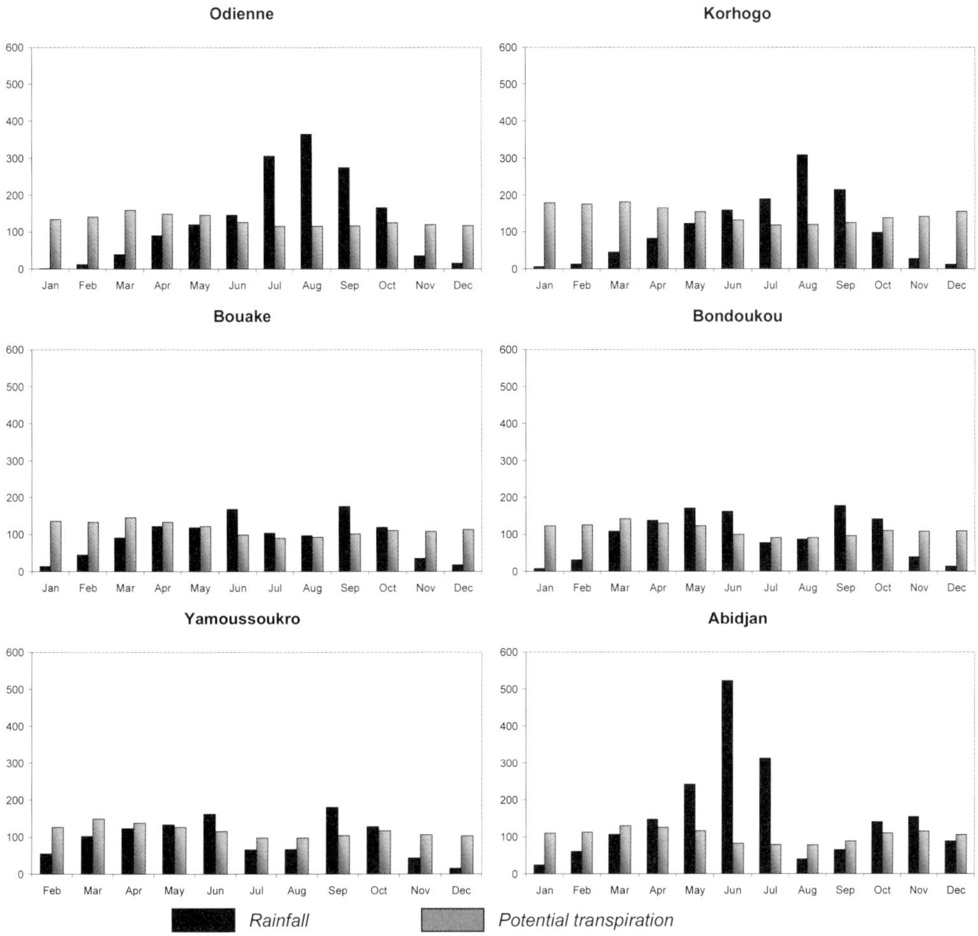

Fig. 9.5 *Côte d'Ivoire: mean rainfall and potential transpiration, mm.*

net rainfall and the records were naturalised by adjusting the flows to restore the effects of reservoir storage and evaporation. This involved the collection of long-term rainfall records and the estimation of basin rainfall series, and the estimation of open water evaporation and potential transpiration for a number of sites. This revealed a transition over the country from a single rainfall series in the north to a bimodal distribution in the south. It was evident that the water balance was affected by seasonal rainfall distribution. Examples are given in Fig. 9.5 of rainfall and potential transpiration to illustrate the distribution over the basin.

A major finding of the hydrological study was that the water resources of the basin had reduced substantially since a previous study in 1966. This decline was compared with flows of the Senegal and Niger during the same period, which showed similar declines. This led to a review of the history of water resources in different parts of Africa over the same period (Sutcliffe & Knott, 1987). It was evident that African river flows are extremely sensitive to variations in rainfall, and that runoff is disproportionately derived from areas with rainfall above 1500 or 2000 mm where runoff exceeds 400 or 1000 mm. These areas are largely confined to the coastal belt of West Africa

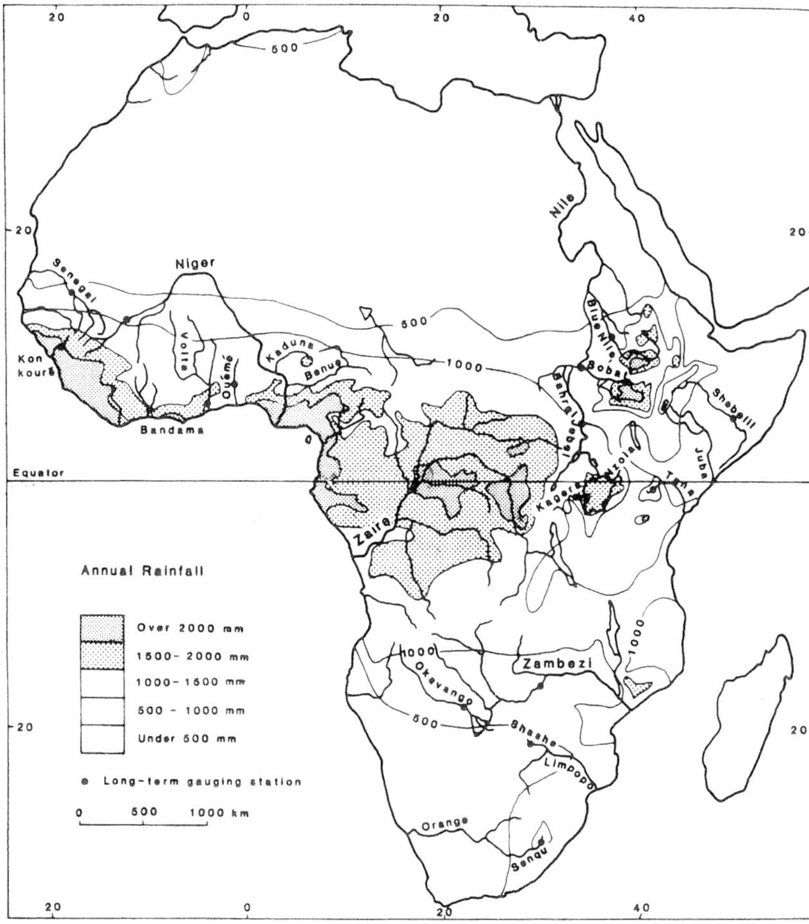

Fig. 9.6 Mean annual rainfall and drainage basins in Africa (from Sutcliffe & Knott, 1987).

and the Congo basin, and to limited mountainous areas in Ethiopia and around Lake Victoria (Fig. 9.6). A number of major rivers in West and East Africa have their sources in these limited areas of high rainfall, and their flows are therefore related. Analysis of the flow records of 16 rivers showed that the decline in river flows over West Africa during the 1970s also spread as far east as the Blue Nile basin. In East Africa, by contrast, the flows of various rivers like the Tana in Kenya, the Shebelli in Somalia, and the outflows from Lake Victoria, had all increased after heavy rains in 1961–1964. There had been variations over long periods in basins further south in Africa, as illustrated by the flows of the Zambezi and other rivers, but the periods of relatively high and low flows did not appear to coincide in the same way as in West and East Africa.

OTHER EXAMPLES OF CLIMATE VARIABILITY

Examples of the effects of variability of climate are not confined to Africa. During a visit to north China in 1989, it was clear that the problems presented by prolonged

Plate 9.1 *Cultivation in a reservoir on the Shantung Peninsula, China.*

drought coinciding with industrial development were serious. It might even have been necessary to consider moving the capital if the problem could not be solved. This situation was illustrated vividly by a visit to a reservoir in the Shantung Peninsula (Plate 9.1) where a farmhouse just upstream of the embankment, in the lowest part of the reservoir, was a base for cultivation, and there was the sight of a large boat grounded in the empty reservoir.

In the Jordan valley, long-term records of rainfall at Jerusalem are available to illustrate climate variations since 1846; periods of drought in the 1920s contrast with other periods. The records correspond reasonably with fluctuations in Dead Sea levels, though recent abstractions have superimposed a falling trend. This area is discussed more fully in Chapter 11.

ANALYSIS OF YIELD

Once a reasonable length of flow record at the project site is available for analysis, it is possible to estimate the yield of the site, with or without reservoir storage. Examples of yield without storage are given, in which water is abstracted on the basis of a run-of-river scheme. The yield can be expressed as a firm yield, or one which would have been met throughout the period of record; alternatively the risk of shortage could be expressed in statistical terms, in terms of the frequency of shortage or of its reciprocal, the return period or the average interval between shortages.

Run-of-river project

The simplest form of yield analysis is for a run-of-river abstraction or hydroelectric power project where additional storage is not intended. In such cases low flow frequency analysis is a common tool to estimate the yield of a run-of-river scheme. For example, without the use of storage within the lake, the firm outflow from Lake

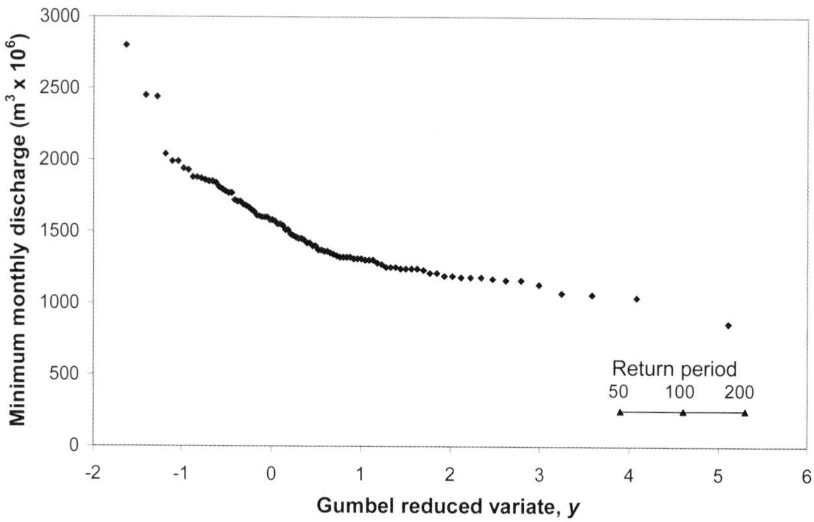

***Fig. 9.8** White Nile at Malakal: annual minimum flows, 1905–1997.*

Victoria at the Owen Falls dam can be taken as the minimum flow over the period of record, which started before 1900. The minimum lake level and outflow occurred in January 1923, when the monthly outflow was 924 $m^3 \times 10^6$, or 345 $m^3 s^{-1}$. More information could be derived from a statistical analysis of the minimum outflow in each year, in order to deduce the frequency of occurrence of a given flow. However, the Jinja flows are not easy to analyse in this way, as the outflows in successive years are linked by lake storage.

An example of low flow frequency analysis on the Nile system was at Malakal, where the White Nile flows combine the outflows from the Sudd, which are linked with the outflows from Lake Victoria but modified by the effects of downstream lakes and the wetlands of the Bahr el Jebel, and the Sobat River which drains a wide area of southwest Ethiopia. The annual minimum series of monthly flows were plotted against Gumbel y according to the Gringorten plotting position. This suggests that there is a limit below which the flow does not go. The flow series is shown in Fig. 9.8, where the significant firm flow derived from the Sudd is evident.

RESERVOIR YIELD

In practice the firm flow of a river is often insufficient to meet the demand, and it is necessary to introduce the concept of reservoir storage to increase the water available. The analysis of yield at a site with a reservoir requires not only a series of river flows but also the dimensions of the reservoir which may be generalised as a relation between reservoir area and volume. There are many ways in which a reservoir scheme may be analysed; in general terms the analysis may be based on historical trials, stochastic generation of synthetic flow series, or statistical analysis of successive years. The following discussion is limited to methods used in successive studies in Botswana, where examples of most types of analysis were used over the years (Parks & Sutcliffe, 1987). As the available data increased, new techniques of analysis made it possible to use more direct methods of assessing the yield of potential dam sites. The methods progressed from the trial of a single critical period, through an analysis based on a generalised study of reservoir geometry and drought frequency, to a selection of techniques of direct reservoir analysis. These included a simple historical trial, an analysis of deficient volumes, and the transition matrix approach, all based on extended flow series.

Rainfall–runoff modelling in Botswana

The extension of flow series in Botswana was discussed in Chapter 8. The water balance of northeast Botswana presents an extreme example of periods of water deficit. In early studies an empirical relation between basin mean rainfall and runoff was used to estimate mean flows at a number of potential reservoir sites. In a later study, sufficient runoff records were available concurrently with rainfall records for the Pitman rainfall–runoff model to be calibrated and used for extension.

With about 15 years of measured flows at the Shashe dam, and rainfall records available back to 1922, the model was used to generate flows which covered 60 years in total. The calibration of the model was based on comparisons of the mean annual flow and its variability, together with the seasonal distribution of runoff. There was considerable variability in the annual runoff (Fig. 8.11).

The simplest method of estimating the yield of a reservoir is the historical trial, in which the reservoir is modelled over the period of flow records. The inflows are added to an initial storage, allowance is made for net evaporation over the reservoir area, and a yield is abstracted from the stored volume. When the reservoir is full, excess water is spilled. When the reservoir is empty the trial fails; the yield is gradually increased until this occurs and this determines the yield of the reservoir over a given period of years. This could be repeated for larger sizes of reservoir to derive a relation between reservoir size and yield. Alternatively, river inflows could be plotted in cumulative form (Fig. 9.9) and lines drawn across the "shoulders" of the graph, using the classical Rippl method to deduce the yield for a given storage volume. The gradient of the line represents the yield and the maximum vertical distance between the line and the graph of cumulative inflow represents the volume of storage required. However, it is not easy

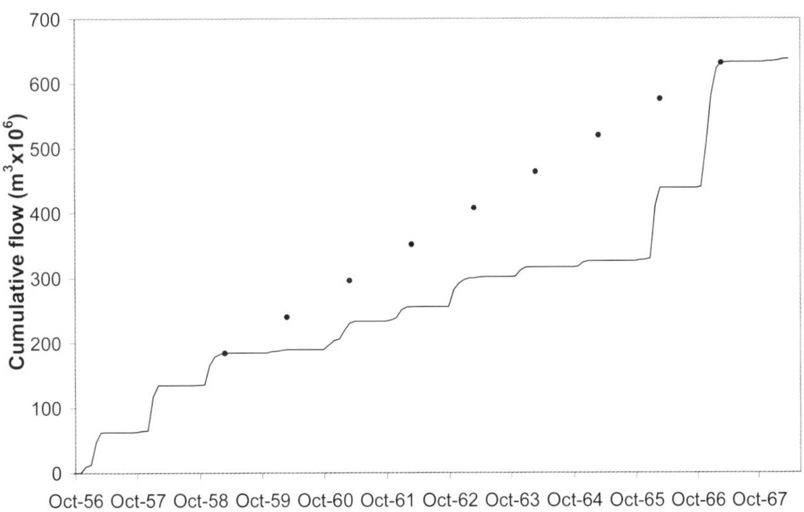

***Fig. 9.9** Rippl diagram: Shashe flows, Botswana.*

to allow for reservoir evaporation, as this varies with the current area of the reservoir; evaporation is a major component of reservoir water balance in arid countries like Botswana. Moreover, this approach makes no estimate of frequency of failure and depends entirely on the characteristics of the flow series, including the order of the sequence.

OTHER METHODS OF RESERVOIR ANALYSIS

Other methods of reservoir analysis provide estimates of frequency of failure to achieve a given yield. A series of analyses in Botswana (Parks & Sutcliffe, 1987) illustrate the different techniques available as the length of flow record increased and analyses progressed from reconnaissance to detailed design. In 1968, when only six years of flow records were available above a reservoir site on the River Shashe, rainfall records indicated that the recent drought of 1963–1965 was the most severe since 1911–1914; in order to test a three-year critical period, a third year of drought was added to the flows recorded in 1963–1965. A simple correlation between Francistown rainfall and river inflow was used to estimate the probable reservoir storage at the beginning of the critical period for various reservoir volumes. These values were compared with the reservoir volumes required to provide a given supply during the critical period to deduce the yield for several reservoir volumes. This provided an estimate of the yield that would have been available from a specific site during a historic drought. Because the drought was within the memory of local people the benefits of the reservoir were easy to convey; however, the uncertainty of the return period of the drought makes an economic analysis rather subjective.

When reservoir yield estimates were required in 1976 at a large number of sites for a preliminary regional feasibility analysis, reservoir yield methods developed by Midgley & Pitman (1969) were used. The method, which was based on the analysis of many flow records in neighbouring South Africa, depends on dimensionless tables of cumulative deficient flows, expressed as percentage of mean annual runoff, for different return periods. They present curves for different climate zones in South Africa, and for example the curve for the zone adjacent to the Botswana border gives for a frequency of once in 50 years a total 8-year cumulative flow of 2.02 times the mean annual runoff. These curves are then combined with generalised reservoir geometry to derive dimensionless relations between net yield and reservoir capacity with both expressed in terms of mean annual runoff. Because the dead storage greatly affects the relation between gross and net yield, which takes account of reservoir evaporation, and because dead storage could not be estimated without site survey, a simple relation was derived to estimate net yield of 1 in 50 year frequency against storage as a percentage of mean annual runoff. This curve was used to estimate the net yield at each site, and 13 out of 75 sites were selected for detailed reservoir survey in the second phase of the study.

The method of estimating net yield was then applied to each individual reservoir. The curve determines the storage required to maintain a given yield, if reservoir evaporation is neglected. However, reservoir evaporation is important in this area and can be deduced from reservoir geometry, which may be expressed as $A = cV^n$ in terms of area and volume, and the reserve or dead storage. The method assumes that the reserve storage is reached at the end of the design drought, and, working backwards in time, the storage volumes and thus areas during the drought are calculated from the balance between inflow, abstraction and evaporation. The net evaporation rate is estimated as the difference between open water evaporation and dry year rainfall. This calculation determines the storage needed to maintain a given net yield, and a series of calculations provides a curve linking net yield to effective storage at each site. This method provided estimates of yield available for different return periods but assumes that the reservoir is full at the beginning of the drought. It was therefore suitable for the reconnaissance survey of 1976 where the yield potential of many sites was required. However, the approach depended on the considerable previous analysis of a large number of flow records, and such regional information would not usually be available.

For subsequent analyses of specific reservoir sites, more detailed behaviour methods were employed. Once a long sequence of runoff records, measured or synthetic, was available, the water balance of a hypothetical reservoir could be modelled to deduce the yield which would have been obtained from a given storage. In the simplest method the inflows were used in a monthly balance model. The number of years or months in which a failure occurred could be expressed as a percentage of the total number of years or months. This provides an estimate of the return period of failure for a particular yield, and a storage–yield curve may be deduced. However, for the periods available of about 60 years this approach is somewhat imprecise.

A new technique based on deficient volumes was developed to relate yields more precisely to associated failure rates. This is also based on reservoir water balance, but in theory incorporates an infinite reservoir volume. The water balance then provides a series of annual drawdowns, or deficient volumes, for statistical analysis. The series of annual maximum monthly deficient volumes is ranked and plotted using a log-normal plotting scheme (Parks & Gustard, 1982). In the absence of evaporation this method can be used directly to determine a yield–storage relationship, but where evaporation is important, as in Botswana, the method can be adapted to define a series of specific storages and calculate the associated deficient volumes. As the drought severity increases, the deficit increases until the net reservoir capacity is reached and a failure occurs; thereafter the deficit is constant (Fig. 9.10). Interpolation from the series of reservoir volumes leads to identification of the volume which supplies the required yield with the design return period of failure. The exercise is repeated with different yields and the results interpolated. The method combines a probabilistic approach with reservoir simulation. It links the yield with the return period of failure for a given reservoir capacity.

The 59-year record also enabled matrix methods of reservoir yield to be applied to the study of Francistown supply. The Gould method, described in McMahon & Mein (1978), was used. A monthly water balance, including evaporation, produces a transition matrix describing the likelihood of ending the year with a given live storage, conditional on the initial year storage. Each year of the inflow data is routed separately through the reservoir, and a transition matrix is derived from the number of times each end of year state is reached from each initial state. The number of occasions of failure and spill are included in the analysis. The steady state probability of storage contents can be derived from successive combination of the transition matrix and the starting conditions; this method takes no account of the order in which annual flows occur. The results can also be expressed as probabilities of failure and water supplied for a given reservoir capacity.

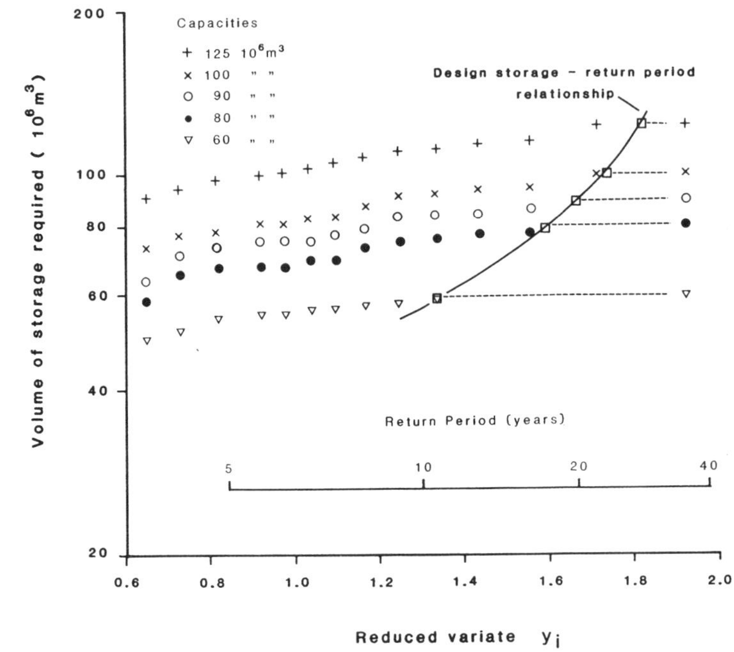

Fig. 9.10 *Botswana: deficient volume analysis (from Parks & Gustard, 1982).*

It is also possible to incorporate the possibility of rationing into reservoir analysis. The return period of failure has historically been used in reservoir design; however it is unrealistic to assume that the reservoir would be operated at full demand until the reservoir fails. Rationing would be used to conserve the available storage when the water level is low. The Gould matrix method can be adjusted to include rationing based on simple design rules within the water balance. Where two or more water uses depend on the same reservoir, it is useful to consider the effect of reducing one form of use. Examples range from garden watering in an arid climate where domestic or industrial water supply is more essential, or irrigation of annual or perennial crops where the area planted with annual crops can be reduced in years when the reservoir storage is low. The combined yield of several reservoirs can be estimated with flow series extended by stochastic generation of flows to provide flow series to test a more complex system.

DOWNSTREAM EFFECT OF RESERVOIRS

Following a number of analyses of individual reservoirs in Botswana, a related problem was studied. A number of small dams, for village supply or livestock watering, were being constructed upstream of major reservoirs. It was not clear how much effect these dams would have on the inflow to the major reservoirs downstream, on which the urban and industrial water supply of the country depended. The problem was studied empirically (Meigh, 1995) by deriving the surface areas of these 320 small dams from aerial photography or maps and surveying a sample in different parts of the country in order to deduce relations of the form $V = aA^b$ between area and volume. The volumes of these dams in each basin were deduced from the measured areas and a typical non-dimensional area–volume curve was derived. The three major catchments draining to the main reservoirs were divided into sub-catchments, and the small dams were divided into categories and amalgamated into composite dams within each sub-catchment. Daily flows for the period 1970–1990 were generated for each sub-catchment using a rainfall–runoff model and the operation of the small reservoirs was modelled, with abstractions estimated according to use for watering or irrigation. The outflows from groups of reservoirs were deduced and channel losses included to estimate inflows to downstream reservoirs while the number and total capacity of the small dams within the basin were increased in the model. The effect of the number and size of the small dams was examined by this modelling; the total capacity of the small dams was found to be the main factor affecting the runoff. Dams with a total capacity of 10% of the mean annual runoff caused a similar decline in catchment runoff. However, dams in the headwaters were found to have a smaller effect, while a few large dams had less effect than a large number of small dams with the same total capacity. The effects on floods and sedimentation were also studied.

COMBINED RESERVOIR ANALYSIS

It is often possible to link the operation of several reservoirs and this can improve the efficiency of the system, or in other words reduce the frequency of failure. An example where several different functions could be combined was investigated on the upper Vistula above Krakow in southern Poland. Several reservoirs had been constructed to provide water supply for urban and industrial use in this area, with some storage allocated for flood control downstream.

In the assessment of the potential role of a reservoir under construction, the contribution which this reservoir could make to flood control was investigated. The inflow hydrographs for a number of flood events were collated, and the effect of flood control operation on flows through Krakow city was investigated by computer-based reservoir trials. Fixed control rules were applied and the reductions of downstream peak flows were noted. It was found that the reductions were improved by combining the roles of the different reservoirs and transferring the volumes dedicated to flood control from one site to another to take account of the potential of the different tributaries. The improvement in frequency of downstream flooding was significant.

Where complex reservoir systems are to be designed or controlled for a variety of objectives, such as water supply, irrigation and hydroelectric power production, linear or dynamic programming may be used to analyse the system in order to maximise the benefit, which has to be expressed for each function in economic terms.

CONCLUSION

The assessment of the water resource availability at a site or a number of sites depends in the first instance on an estimate of the mean annual flow and its variability, both seasonally and from year to year. Because of the variations which can occur in even the mean annual flow over time, especially where runoff depths are low as in Africa, it is important that estimates should be based on as long a period as possible. This period may be increased by extension of river flows by correlation with longer flow records or rainfall records. Such extension is simpler in humid than in arid climates, as evaporation losses are more predictable. Monsoon climates may be treated as seasonally humid. Flow records may also be extended by stochastic generation of flows with similar properties to the measured record.

Once long flow records have been obtained or estimated, the water resources yield of a site may be based on direct abstraction or on the inclusion of a storage reservoir. The yield of a run-of-river scheme may be assessed by statistical analysis of annual minimum low flows. However, many sites require the provision of a reservoir to maintain an adequate yield, and this will involve reservoir analysis to relate reservoir capacity to the yield which can be maintained with a given frequency of shortage. There are many available forms of analysis, and the choice depends partly on the length of records available; examples from Botswana include historical trials, statistical analysis of deficient volumes and transition matrix methods. Multiple objectives and the effect of rationing control can be included. Whereas such techniques would at one time have required analysis on mainframe computers, they can now be carried out using software available for a PC. The spreadsheet is now a key tool for those designing water resources projects.

10 FLOOD ESTIMATION

The estimation of flood potential at a project site differs from the study of other aspects of the hydrological cycle. Most hydrological studies are based on the analysis of measurements within a basin and the deduction of typical values of different processes like basin rainfall, evaporation or runoff, leading to a water balance tying these elements together (Chapter 8). The study of floods involves the extrapolation from flows experienced at a site or at comparable sites, to floods which can be expected to occur with a given risk or frequency of occurrence, or even to a flood magnitude which is not expected to be exceeded. Thus most other studies can be based on what has been observed over a basin, but flood estimation involves an attempt to deduce what might occur outside the limits of what has occurred in the past. It might seem, in relation to the title of this book, that we are dealing with a question of imbalance, but floods are in fact closely related to a balance between rainfall and soil moisture deficit. Several possible approaches to flood estimation are described, with examples drawn from different environments around the world.

REQUIREMENTS FOR FLOOD ESTIMATION

Flood estimates are required for a wide variety of civil engineering projects, and almost all projects situated near a river need a flood estimate of some form. At its simplest any structure will need an estimate of the risk that the river level will exceed a certain elevation in order to design protection works. Any structure designed to pass the flow of a river—a culvert, a bridge or a channel reach with flood protection levees—will be based on a flood-frequency relation and an appropriate frequency of the design flow being exceeded. Major structures, like a dam and its discharge spillway, may be designed to pass the flow of a lower frequency or longer return period; the return period is the average interval between exceedances of a given flood, or the reciprocal of the frequency. It may, however, be designed to pass an estimated maximum flood to ensure protection of the structure itself or downstream property against dam failure.

Thus flood estimates may range from flows of low return periods to much rarer events or even to the likely maximum flood. It may be necessary for some purposes to express the flood in terms of river level instead of flow. Because rare floods, by definition, are unlikely to have occurred within a short period of record at a single site, it is essential to enlarge the record on which the estimate is based. This may be done in several ways: information may be sought on historical floods which occurred before formal records were begun; records may be included from other sites within an area which could be considered to have a similar flood regime; rare floods may be assessed by combining physical or statistical estimates of rare storms with a means of converting storm magnitude and distribution in time and space over the basin to a flood hydrograph at the project site.

Thus there are a number of ways in which the flood estimate may be made, and the choice may depend on the information available and the design return period. The choice of methods is discussed first, and then the different methods are summarised with examples from different regions.

CHOICE OF ESTIMATION METHODS

Various methods are available for estimating flood magnitudes, and they have different data requirements. For example, the *Flood Studies Report* (NERC, 1975) presented a number of techniques which were considered appropriate for estimation in the British Isles. At the time, hydrological data in Britain were largely collected to estimate water resources, and there was no central archive of flood flows. Before existing techniques could be tested and developed, the hydrological team involved in the study, based at the Institute of Hydrology, Wallingford, spent a year visiting all the gauging authorities and some 1200 gauging stations; station reports were prepared and flood rating curves discussed with the local authorities and their reliability graded. Some stations were found to lack accurate high flows and others were at the time too short to be useful. Collection and microfilming of about half a million river level charts were followed by extraction of peak flow records at some 500 gauging stations for statistical analysis, and the extraction of flow hydrographs from 140 sites for unit hydrograph studies. Daily rainfall data, recording raingauge charts, and soil moisture statistics were also collected for comparison with flood hydrographs. Basin characteristics were derived from maps for all the sites suitable for analysis. Thus recommendations were based on a very large body of data on storm rainfall and measured river flows, and on the analysis carried out on these data to obtain relations between observed rainfall and subsequent runoff, and also between observed flows and basin characteristics. Although the empirical relations, for example between mean annual flood and basin characteristics, apply to the area for which they were derived and have been revised by subsequent analysis, there is no reason to believe that the principles involved and the criteria for choice of method are not still valid.

In 1999, Reed *et al.* produced the *Flood Estimation Handbook* which brought up to date the earlier *Flood Studies Report*. Application of the recommended methods is based on analogue records rather than geographical grouping and requires a CD-based set of basin descriptors, which limits its direct use to the UK.

The *Flood Studies Report* provided a choice of means (Fig. 10.1) of estimating the flood hydrograph of any return period, or the estimated maximum flood, from an estimate of the storm depth and its time profile and its conversion to river flow. It also provided a means of estimating the flood peak of any return period by statistical methods, either from flood series measured at the site or, if these were not available, from basin characteristics derived from maps. The choice between these two main approaches, based either on analysis of rainfall and runoff, or on statistical analysis of flood series, and the selection of methods of flood estimation suitable for different circumstances, have been discussed by Sutcliffe (1975, 1978). One method may have advantages over another in certain circumstances, but it may also be desirable to compare and combine more than one method of estimation. The first choice between unit hydrograph analysis of the flood corresponding to a design storm, and direct statistical analysis of peak flows, depends largely on two questions:

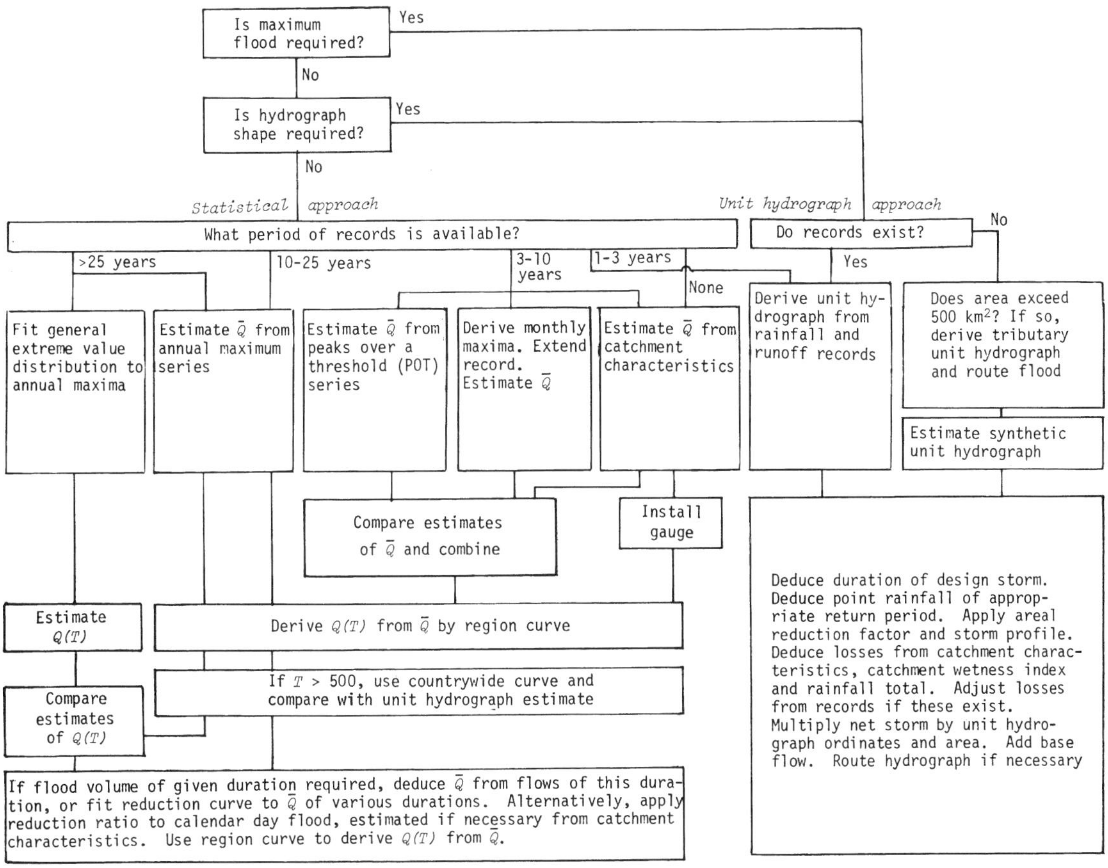

Fig. 10.1 *Choice of flood estimation methods (after NERC, 1975).*

- Is the design flood hydrograph required in addition to the peak flow?
- Is an estimate of the maximum flood required rather than the flood of a given return period?

If the answer to either of these two questions is yes, then the rainfall–runoff approach becomes necessary, of which the unit hydrograph method is an example; if not, then either the statistical approach or the rainfall–runoff approach is suitable, and either or both may be used. The choice also depends on whether there are long, short or no records at or near the site where an estimate is required.

A summary of the recommendation is included in Fig. 10.1. If the flood $Q(T)$ of a given return period T is required and a flood hydrograph is not needed, the statistical approach provides the simpler technique. The statistical approach depends on an estimate of the typical annual flood at the site, based on the mean annual flood or the median flood, which may be estimated from the series of flows at the site. The index flood may be derived from the series of maximum peak flows in each year of record. Where no records exist at a site the mean annual flood can be estimated from basin characteristics; useful characteristics in the British Isles included area, stream frequency, channel slope, net short-term rainfall and indices of soil and lake-affected fraction. These can be deduced from standard topographic maps or from thematic maps illustrating soils or rainfall for the study area. The relation between basin characteristics and the index flood can be deduced by analysis of the floods and basin characteristics of stations within a region or group of sites with similar characteristics. However, the standard error of estimate of such an estimate is likely to be high, and it is preferable to base an estimate on flows measured at the site; for this reason, it is essential to install a gauging station at a site where an important project is foreseen.

If the period of record available is short, the index flood estimate may be improved by extending the record using adjacent long-term records. A "peaks over threshold" analysis may also prove useful; this approach extracts all independent peaks above a given threshold, using a simple rule for independence. The mean annual flood may be simply estimated from the average number of exceedances per year and the average magnitude of the peaks over the threshold.

Where a longer period of records exists, say over 10 years, the mean annual flood may be estimated directly from the maximum flows in each year of record. The flood of return period T may be estimated from the mean annual flood using a dimensionless growth curve, in which the flood of a given frequency or return period is expressed as a multiple of the mean annual flood. This approach is useful when the records from a number of stations are combined to derive a group growth curve. In the 1975 *Flood Studies Report* these group curves were derived from the stations within a region, but they may also be derived from a set of stations with similar area, rainfall and soil characteristics. For moderate return periods, for a site with reasonably long records, it is possible to fit a statistical distribution to the annual maximum flows, but a group curve gives added stability to the result.

Where there is a need to specify the flood hydrograph, or to estimate the maximum flood, the unit hydrograph approach (e.g. Shaw, 1983) is appropriate. The unit hydrograph is the response of a basin to a net storm of a given magnitude, originally an inch but later 1 cm, in a given time period normally taken as 1 hour; this response is assumed to be constant for a given basin. The unit hydrograph for the project site should preferably be derived from rainfall and runoff records, but may also be estimated from basin characteristics where no records exist. The appropriate duration of the design storm is derived from the unit hydrograph dimensions, and the return period of the design storm is deduced from the required return period of the design flood. The rainfall depth corresponding to this duration and return period may be derived, either at a point or over a basin, from the meteorological study, together with a time profile of this storm. The mean point rainfall of the given duration and 5-year return period can be deduced from a map; a growth factor is applied to give the mean point rainfall of the appropriate return period, and an areal reduction factor is applied to give the basin rainfall, which is distributed in time by a standard profile.

The proportion of the storm giving rise to immediate runoff is estimated from the basin soil and urban fraction, taking account of appropriate antecedent conditions and rainfall depth. The net storm profile should be multiplied by the unit hydrograph and base flow added, to give the design flood. This procedure may also be used to estimate the maximum flood, using an estimate of the maximum storm rainfall and also more conservative assumptions about antecedent conditions, storm profile and unit hydrograph; the unit hydrograph can be adjusted by reducing its time to peak and increasing its peak. An allowance for snowmelt runoff is added where appropriate.

Thus there is a wide variety of methods available for estimating floods, and the choice depends on data availability and to some extent on the circumstances of the estimate. It is also desirable to use alternative methods and compare the results.

STATISTICAL DISTRIBUTIONS

The fitting of statistical distributions to annual maximum flood values at a single site was the subject of considerable research, following the adaptation of extreme value theory by Gumbel (1941) to the flood problem. He apparently met an engineer on a skiing holiday and described his research in theoretical statistics, based on work by Fisher & Tippett, and was encouraged to apply this work to flood problems. He noted that the distribution of the largest values in sets of numbers drawn from different parent distributions converged to one of three distributions, which became known as Gumbel Types I, II and III. He proposed the double exponential or Gumbel Type I distribution for annual maximum floods. Bernier (1956) drew attention to flood records which conformed to the Fréchet distribution, which was equivalent to Gumbel Type II. When plotted on specific graph paper, which made use of a "reduced variate" linked with frequency, these three curves emerged as straight lines, concave upwards and concave downwards respectively. Gumbel suggested that the first two distributions were appropriate to floods, while the third was more appropriate to the distribution of low flows or droughts.

In 1955, Jenkinson showed that the three distributions were special cases of a General Extreme Value (GEV) distribution, and this has been widely used after its adoption by the *Flood Studies Report* (1975) on the grounds that it had some theoretical basis and was flexible in application. The GEV distribution is defined by:

$$F(x) = \exp\{-[1 - k(x - u)/\alpha]^{-1/k}\}$$

and is equivalent to Gumbel Types I, II and III as $k = 0$, $k < 0$ and $k > 0$.

However, the fitting of any distribution to a single station record is subject to a high degree of uncertainty, particularly as the typical return period for which a flood estimate is required will usually greatly exceed the length of record available. Although the choice of distribution will affect the extrapolation of the flood record markedly, as the choice between the three Gumbel types would illustrate, the stability of the curve can be improved by increasing the volume of records on which the fitting depends. There are various ways of extending the flood record at a site. In some cases it may be possible to make use of historic records, which are often available in the form of levels from flood marks or local histories. However, a more general way of increasing the effective length of flood records is to include the records of similar stations, and by making the records dimensionless to analyse the whole set as though they were from a single site.

In the investigations leading to the *Flood Studies Report*, the grouping of stations was based on geographical regions. This was partly chosen for simplicity, but it was also based on the belief that the main factor governing the shape of the dimensionless flood frequency curve was the rainfall of the basin, and that this had a geographical basis. The statistical studies showed that the variability of floods was linked to climate indices more closely than to size or soil factors. In order to derive a rainfall factor more directly related to the flood-producing role, it seemed clearly preferable to use a short-term rainfall index than the annual average rainfall used in earlier studies. Further, the effect of climate in determining the proportion of rainfall which leads to direct runoff was believed to be based on soil moisture state, and the concept of one-day rainfall less mean effective soil moisture deficit was developed for this purpose (see Chapter 5).

The growth curves for the average rainfall for each of the flood regions used, when adjusted for the mean soil moisture deficit to give net rainfall, were very similar to the flood growth curves for the same regions. During the research leading to the 1975 report, the mean climate characteristics for all gauging stations within each region were compiled; these were *RSMD* or net 1-day rainfall of 5-year return period, *SMDBAR* or mean effective soil moisture deficit, and the rainfall growth factor (*MT*/*M5*) as a function of 5-year daily rainfall. These regional characteristics were used to obtain growth curves of the *RSMD* or net rainfall of different return periods. These values are illustrated in Fig. 10.2; the strong similarity between the two sets of growth curves is evident.

It was possible to interpret the statistical relation in physical terms. Although the relation between mean annual flood, *Qbar*, and basin characteristics appears complex and somewhat arbitrary:

$$Qbar = 0.021\ AREA^{0.94}\ STMFRQ^{0.27}\ S1085^{0.16}\ SOIL^{1.23}\ RSMD^{1.03}\ (1 + LAKE)^{-0.85}$$

it can be argued that the equation shows the mean annual flood as proportional to basin area (*AREA*) and net rainfall (*RSMD*), and that the soil term (*SOIL*) represents percentage immediate runoff, while the terms for channel slope (*S1085*) and stream frequency (*STMFRQ*) reflect the shape of the basin response or unit hydrograph.

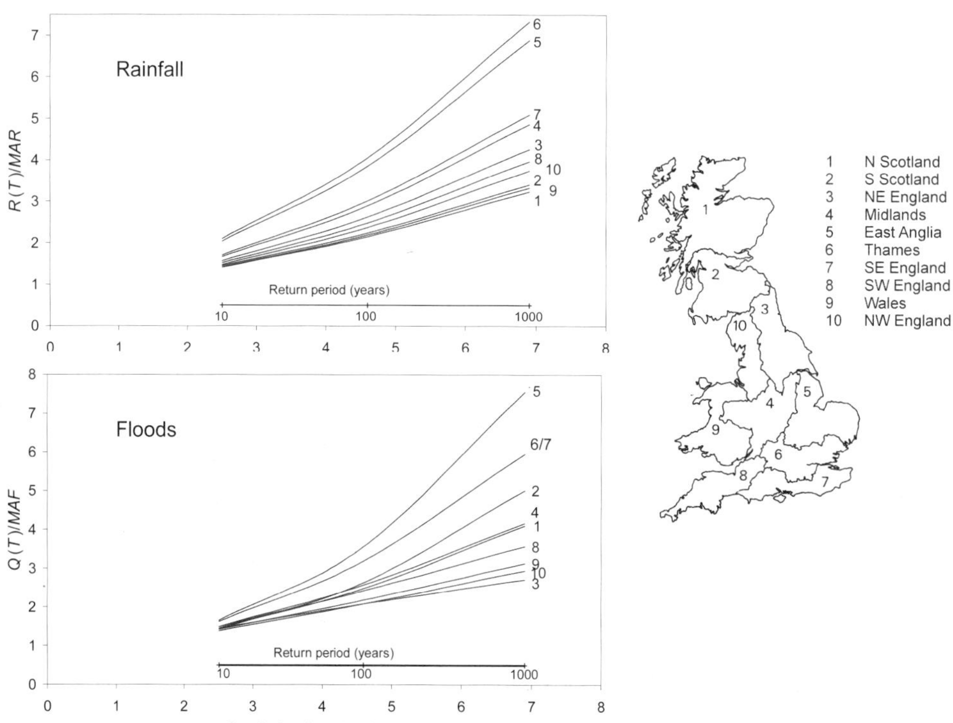

Fig. 10.2 *Growth curves of net rainfall and floods in Britain: the curve numbers refer to the regions shown on the map.*

Subsequent research (e.g. Wiltshire, 1986; Reed, 1999) has led to the use of sets of gauging sites, which are similar to the project site in terms of basin area, rainfall and soil type. However, in parts of the world where thematic maps are rare the concept of geographical regions is a useful starting point for flood studies.

FLOOD STUDIES IN DIFFERENT AREAS OF THE WORLD

The completion of the *Flood Studies Report* naturally led to the question of the relevance of the research to other countries. It was felt that the general approach was relevant: the emphasis on appraisal and collation of flood records; the choice between methods of estimation, and the circumstances in which the statistical approach or the deterministic approach is appropriate; the regional approach, in which the mean annual flood is estimated from records and the growth curve used to convert to floods of a given return period; the concept of preserving the return period of the design storm through the conversion to a flood hydrograph.

One immediate application of the techniques developed was the regional growth curve, which was used in a number of overseas applications after the completion of the report. Experience in different areas drew attention to the very wide variety of flood regimes in different climates. Studies of floods in Botswana, where flood runoff approaches most nearly the traditional concept of runoff resulting from heavy rainfall of high intensity exceeding the infiltration rate of the soil surface, showed that flood growth curves were extremely steep and curved upward. Floods in several West African countries, by contrast, showed an opposite tendency, with flood peaks growing very slowly with return period, to an extent that the fitting of GEV distributions to flood records produced curves which were concave downwards and thus implied that an upper bound existed.

During a study of the water resources of Guinea in 1982, preliminary estimates of spillway design floods were required for comparative economic appraisal of potential sites for hydroelectric power. Daily flow series were available for nearly 20 gauging stations, and annual maximum series were extracted to derive regional flood curves for three provincial groups of stations. There appeared to be no differences between the groups, so that a common curve was used to provide growth factors to convert estimates of the mean annual flood to floods of given return period. Figure 10.3 illustrates this joint curve, which is extremely flat and even appears to curve downwards; however, a straight line relation was adopted as the concept of an upper bound was not accepted.

In a regional study of dam sites in Botswana in 1986, the number of gauging stations with flood records was low, but a regional flood growth curve was derived to support estimates of the mean annual flood at key sites. The annual maxima from 10 stations with a total of 135 station-years of data were analysed using the GEV distribution to give the regional curve also shown in Fig. 10.3; the contrast with the regional curve from Guinea is great. This curve was supported by analysis of corresponding records from South Africa and Zimbabwe.

This experience of contrasting flood curves in different African countries led to a series of investigations in a large number of countries. Although attempts were also made to relate mean annual floods to basin characteristics, these were hampered by the lack of basic data except basin area and annual rainfall; although the results could be useful in a local context, the more generally interesting results involved flood growth curves.

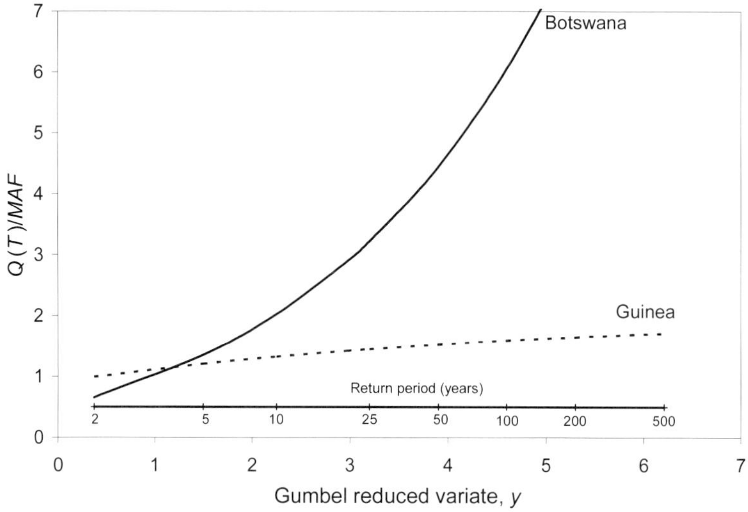

***Fig. 10.3** Growth curves: Guinea and Botswana.*

Semiarid areas

A group of areas with similar growth curves was found in warm semiarid regions, defined as those with annual average rainfall below 600 mm. Flood records were collected (Farquharson *et al.*, 1992) from Botswana and parts of South Africa, Saudi Arabia and Yemen, southern Queensland, parts of southwest USA, and southwestern parts of the former Soviet Union. The records from these groups of stations were used to obtain flood growth curves by fitting a GEV distribution by the method of probability-weighted moments (PWM), which was shown to be an efficient method of fitting frequency curves (Hosking *et al.*, 1984). The Gringorten plotting position was used. The results showed a close fit (Fig. 10.4) between regions of highly diverse geographical location, which shared in common the characteristic of low rainfall.

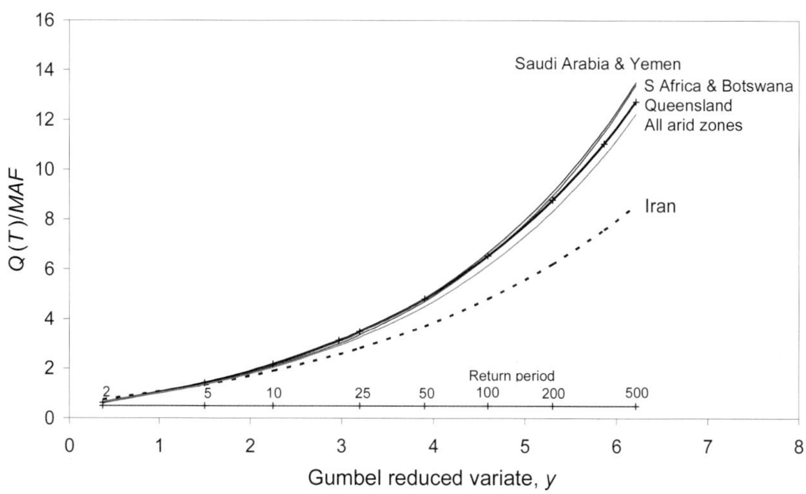

***Fig. 10.4** Growth curves: semiarid regions (after Farquharson et al., 1992).*

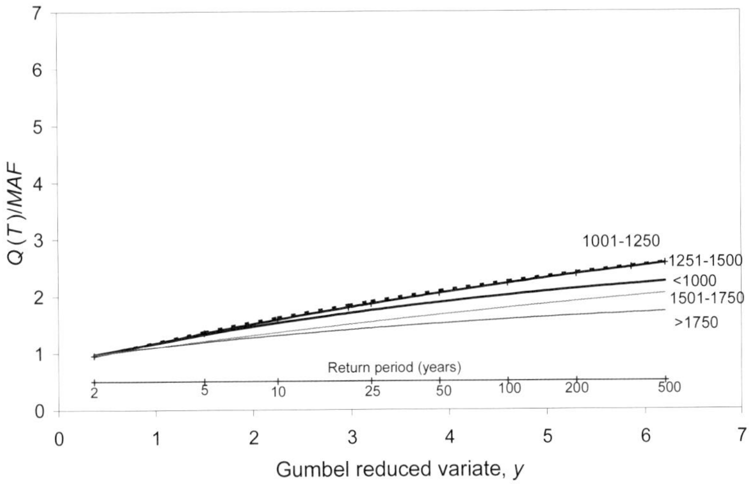

Fig. 10.5 *Growth curves for rainfall zones in West Africa (rainfall ranges in mm) (from Sutcliffe & Farquharson, 1998).*

West Africa

A detailed study of flood records over a wide area of West Africa was made possible by the valuable records published largely in ORSTOM monographs. Flood records from Senegal to Cameroun (Farquharson *et al.*, 1993) were assembled and their mean annual flood values compared with area and mean annual rainfall. Their growth curves were also derived and grouped according to variables which appeared to give similar curves. These variables involved a mixture of geographical regions, defined by latitude and longitude, and also ranges of rainfall (Fig. 10.5) and basin size. However, the growth curves were characterised by their similarities rather than their differences, and they were uniformly of gentle slope and either straight or even curved downwards.

Other regions

A few exceptions to these flat growth curves were encountered in countries with similar annual rainfall. In Kenya, and even more clearly in Sri Lanka, the growth curves were comparatively steep and curved. One difference between the climate of these countries and those of the West African examples is that there is more than one rainfall season, and as a result the dates of occurrence of the annual maxima range throughout the year, while in West Africa the maximum flood dates are extremely regular. It appears that in the West African case the floods are caused by the accumulation of soil moisture and rainfall over a single monsoon season, and the flood is related more to a seasonal rainfall total than to individual storms.

A similar effect is observed in the case of snowmelt rivers located in the northwest USA, Canada and northern Russia; here the growth curves are extremely flat, and the dates of annual maxima again show little variation from year to year.

A preliminary study of flood regimes over Europe (Sutcliffe & Farquharson, 1998) was based on available records, which varied from abundant in western Europe to sparse in parts of central and eastern Europe. This difference reflected the ready availability of published sources rather than the density of gauging stations. A series of curves (Fig. 10.6) indicated an interesting pattern ranging from flat curves in the

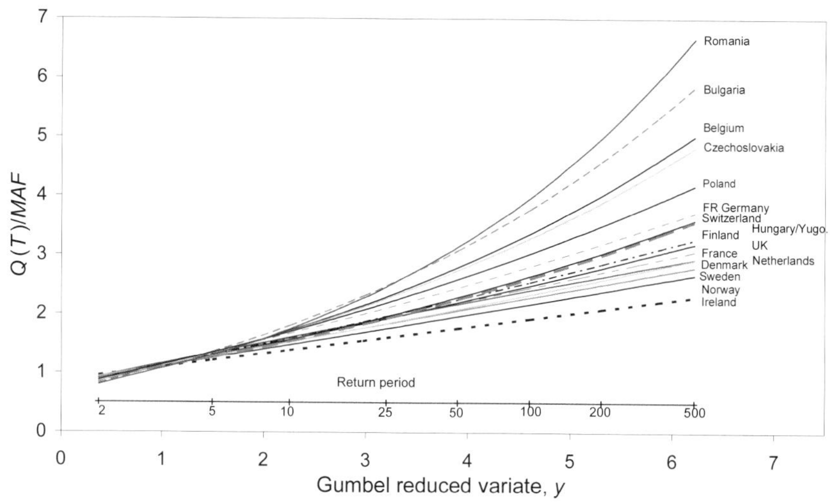

Fig. 10.6 Growth curves: Europe (from Sutcliffe & Farquharson, 1998).

maritime regions of western Europe (e.g. Ireland and Norway) and increasing in gradient and curvature towards the records available in Poland and the Balkans. This study provides material for further research.

Software

Software is available to analyse flood records in various forms. Software linked to the *Flood Estimation Handbook* is relevant to the estimation of floods in the United Kingdom. On the other hand, the *IH Floods* package (CEH, 2001) has been prepared with funding from the UK Department for International Development (DFID) for use in developing countries and is freely available from Wallingford. This program is designed for the storage and analysis of annual maximum flood data. It includes the ability to fit a variety of frequency distributions to the data at a single site. It also allows the selection of groups of stations by a number of different approaches, and the fitting of regional flood frequency distributions to the grouped data. It can thus be used to develop regional flood frequency curves which can be used to make flood estimates at both gauged and ungauged sites in a region.

Polish floods

An example of the use of regional analysis in an apparently homogeneous region of moderate size is given by the investigation of floods in the Krakow region referred to in the previous chapter. The flood regime of the Vistula through Krakow depends on several southern tributaries, which drain the steep Carpathian mountains with average annual rainfall ranging from 700 mm to over 1200 mm. The rainfall distribution is highly seasonal, concentrated between May and September; in addition to snowmelt floods, the flood problem is compounded by summer storms which often track from west to east, so that tributary contributions have a tendency to build up along the main river above Krakow. In a study of the role which a group of dams on the upper Vistula could play in flood control, a regional analysis of flow records above Krakow was

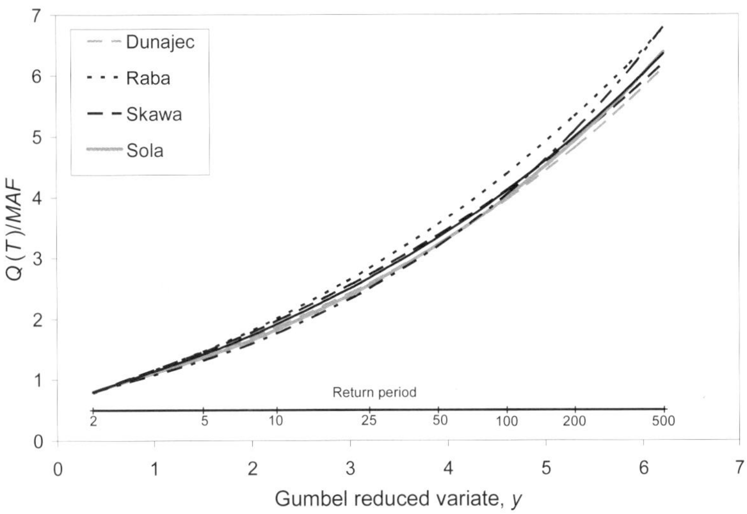

Fig. 10.7 *Growth curves of southern Vistula tributaries near Krakow, Poland.*

carried out. The contributions of the various tributaries to the flood potential were examined statistically. The flood records from the right bank tributaries were assembled and provided 925 years of record from 21 stations. These were analysed in the usual way, and revealed a set of growth curves which were highly consistent but also highly curved (Fig. 10.7). This seemed appropriate as the paper by Fréchet (1927) which led to the proposed use of the Gumbel Type II distribution (Bernier, 1956) was originally published in Krakow.

The effect of this regime is illustrated by flood marks in the city near the river (Plate 10.1), which mark the levels reached in February 1876 and August 1903, together with the earlier higher flood of 1813. The same building is shown in a view (Plate 10.2) of the subsequent flood embankments. After the flood contributions of the individual tributaries of the upper Vistula had been analysed, the potential effect of joint flood control using a combined group of reservoirs was studied by trials involving specific flood events, and it was found that a distribution of the flood control role between the reservoirs would be advantageous. By this means a significant potential reduction in flooding risk could be established.

Hydrological flood forecasting

This Polish example illustrates the benefit of flood forecasting. If a system of meteorological observation or forecasting can be linked to a runoff forecasting system, which was under development, providing estimates of flood hydrographs in time for action to be taken, then it may well be possible to reduce the impact of heavy storms and subsequent runoff by operating reservoirs to attenuate the flood peak. A system of flood warning may at the very least make it possible to reduce the damage or loss of life. The establishment of such a forecasting system may involve radar observation of storm movement, the linking of instrumental measurements of precipitation or river levels and flows by telemetering to a central control office, and the modelling of runoff using updated measurements or modelling of antecedent conditions on the basin.

Plate 10.1 *Flood marks in Krakow showing 1813 (highest), February 1876 and August 1903 flood levels.*

Plate 10.2 *Flood defences in Krakow. The building in Plate 10.1 is the central one in this photograph.*

USE OF HISTORICAL INFORMATION

The inclusion of historical information in flood frequency analysis has the potential to increase significantly the period over which floods are sampled and thus reduce the extrapolation required. Provided that correct statistical procedures are followed this inclusion can improve the confidence with which rare floods are estimated. The historical information can take a number of forms and examples are discussed from different areas. The basic information is likely to be a peak water level, which can be converted to a flow, and a date, which may or may not be precise.

Britain

In Britain historic floods are often recorded on marks or plaques, especially in older cities. During the investigations for the *Flood Studies Report* (1975) an attempt was made to collate all historical information available at or near existing gauging stations, which in most cases had already been assembled by the gauging authorities from physical marks, newspapers, books and verbal accounts. This collection of historical information is presently being continued by the Centre for Ecology and Hydrology (Black & Law, 2004). If a series of historic flood marks is reasonably continuous, it should be possible to assume that all floods above some threshold have been recorded, and the series may be treated as the highest, second and so on over the whole period of record. If any recent floods have exceeded this threshold, their rank may be deduced over the historic and recent records. It may be possible also to take account of a given flood having been the highest since an earlier date or within living memory.

For example, annual maximum floods were available for the Trent at Trent Bridge in Nottingham for the period 1884–1968. There were also flow estimates for the peak floods of 1795 and 1875, with the additional information that the flood of 1795 was reported as the most severe of the 18th century and the lower 1875 flood was the most severe of the 19th century. It could therefore be deduced that the 1795 flood was the highest since 1700, with the 1875 flood probably the second highest. Graphical presentation of historical information illustrates the value of the record, which depends on a reasonably stable channel cross-section. Where a continuous record of annual maximum floods is ranked in size order, each flood can be allocated a plotting position according to its rank i in the N years of record. Cunnane (NERC, 1975; Cunnane, 1978) showed that the Gringorten formula for frequency:

$$F_i = (i - 0.44)/(N + 0.12)$$

gives a reasonably unbiased plotting position based on the Gumbel reduced variate y obtained from:

$$y = -\ln(-\ln F_i).$$

Equivalent plotting positions can be deduced for historic floods, by allocating ranks over various periods. The historic floods for the Trent at Nottingham are included in Fig. 10.8, with the 1795 flood assumed to be the highest since 1700.

China

An even longer example is presented by the Chang Jiang (Yangtze) at Yichang, where annual maximum floods are available for the period 1877–1981, excluding 1979–1980 (Rodier & Roche, 1984). The mean annual flood from this record is 51 700 $m^3 s^{-1}$. An

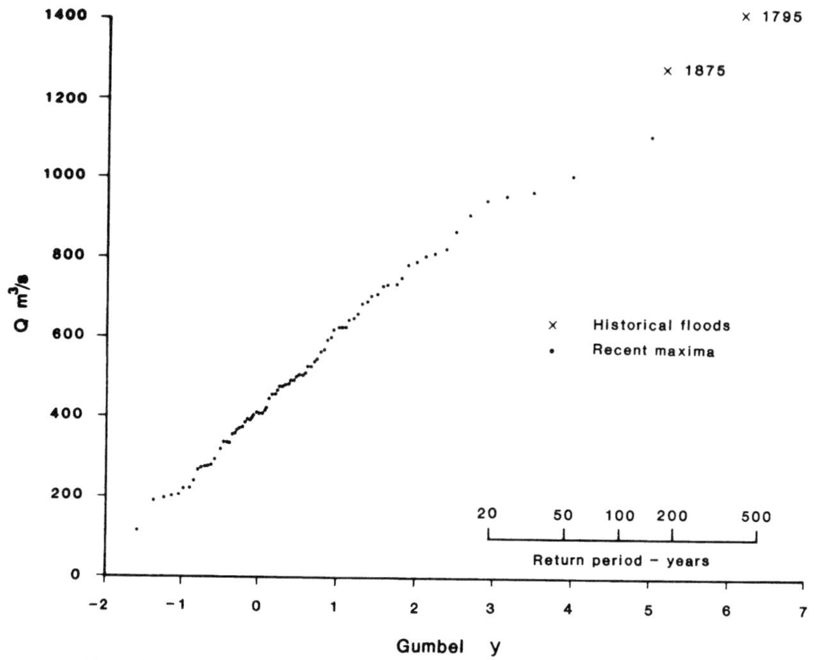

Fig. 10.8 *Historic flood frequency: Trent at Nottingham, 1884–1968 (from Sutcliffe, 1987).*

additional eight historical flood marks are available from 1153 to 1870 AD, and because the rating curve on this reach should be stable the flows have been deduced. Because three floods in 1227, 1560 and 1870 are higher than that of 1153, it is reasonable to assume that the four highest floods since 1153 have been recorded; as the 1860 event was little below the 1153 flood, these floods are included in Fig. 10.9 as the five highest since 1153.

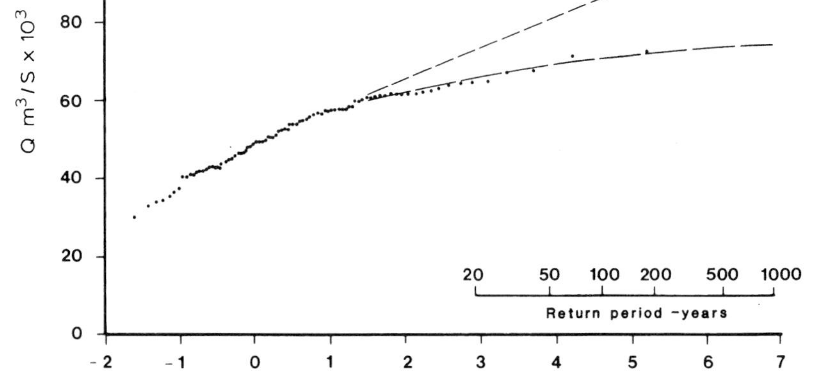

Fig. 10.9 *Historic flood frequency: Chang Jiang at Yichang, 1153–1981 (from Sutcliffe, 1987).*

It is possible to analyse annual maximum series including historic floods by treating the whole record as censored below a threshold during the historic period. The method of maximum likelihood can be used with a censored sample. It can therefore be used to estimate the flood frequency distribution from the recent and historic records. A simple program was used to estimate the GEV parameters which maximise the likelihood for the 1877–1981 record and repeated with the historic records included to give the alternative results plotted on Fig. 10.9. Whereas the recent data suggest a distribution with an upper bound of 77 000 $m^3 s^{-1}$, the addition of the historic floods, with a number of floods above this upper bound, alters the fitted distribution dramatically. The estimated floods for higher return periods are increased by over 30% to correspond with the historic floods.

Indus

By contrast, another example illustrates an exceptional event which it is difficult to incorporate into a statistical analysis. There is a long and apparently consistent record on the Indus at Attock in Pakistan where a staff gauge was installed in 1868 to provide telegraphed flood warning downstream. The channel at the site is controlled by a rock bar which provides a stable control, and the level series can be converted to floods using a single rating curve, whose stability was confirmed by a comparison of rainfall and runoff. The 1868–1978 flood record (Rodier & Roche, 1984) provides an apparently consistent flood frequency relation (Fig. 10.10) from which the prediction of rare events appears simple. However, there is evidence that an exceptional flood occurred in 1841, after a landslide from Nanga Parbat blocked the gorge below Gilgit for six months. Unusually low flows had been experienced downstream during the spring so that the main river had been fordable. A lake "35 miles (56 km) in length and

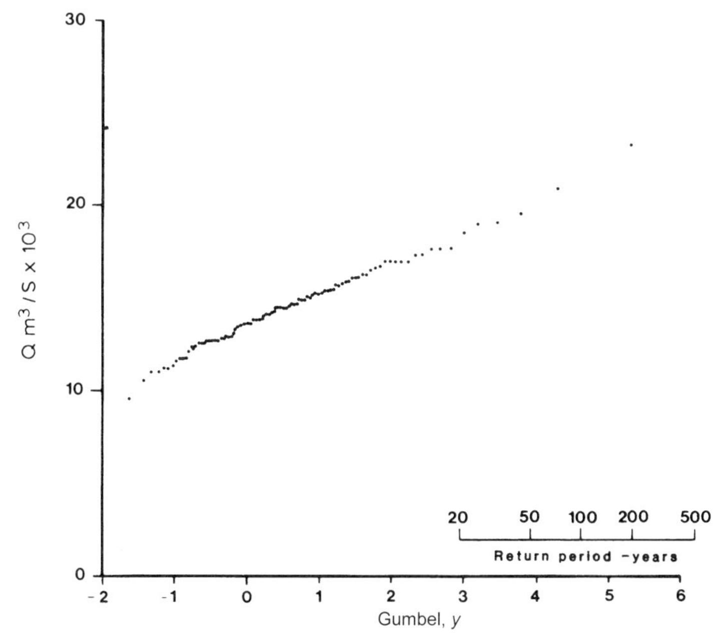

Fig. 10.10 *Historic flood frequency: Indus at Attock, 1868–1978 (from Sutcliffe, 1987).*

of great depth" (Knight, 1893) is said to have drained in one day after breaching the dam. Drew (1875) estimates the length at 35 miles, an average width of a mile, and a depth of about 300 feet above the ordinary river level at its lower end. An account of the subsequent flood described how the river "in a few minutes tore and swept away everything in its course, including a whole Sikh army that had encamped on its banks below Attock with its tents, baggage and artillery" (Wadia, 1975). In the words of the Zemindar of Tarbela (*J. Asiatic Society of Bengal* XVII, 1848, p.232): "As a woman with a wet towel sweeps away a legion of ants, so the river blotted away the army of the Raja". Estimates have been made of the volumes of other floods (Hewitt, 1982), and while it should be possible to estimate the volume stored in this event from satellite imagery, it would not be easy to translate this to a peak flow. However, a tentative estimate of its volume from Drew's information (56 km × 1.6 km × 91 m × 0.5) allowing for the attenuation of depth, gives a volume of 4.08 km^3, or 47 000 m^3 s^{-1} for a day; it is clear that the flood was significantly greater than other floods experienced at this site.

It is also difficult to see how statistical analysis of the flood series can be adjusted to incorporate such an exceptional flood caused by a totally different cause from the other flood events. In fact design estimates for the Tarbela dam above the Attock site were based on maximisation and physical reasoning. A maximum snowmelt flood, derived from statistical analysis, was combined with an extreme storm over the foothills, transformed to flows at the dam site by unit hydrograph methods; the breaching of a natural dam was added to the other flows to provide a design flood.

ESTIMATION OF DESIGN FLOOD

An example of a combined statistical analysis of rainfall and unit hydrograph assessment of runoff was used to estimate a design flood for Gaborone dam in Botswana, where flow records had to be deduced from reservoir levels which covered a short period. Longer daily rainfall records were available at five sites which were considered reasonably independent. Their combined records were used to estimate daily maxima for return periods from 20 years up to 500 years, and locally derived areal reduction factors were used to give mean daily rainfall for the basin of 4300 km^2. Falls of other durations were also deduced from regional studies. In order to derive a unit hydrograph for the site, inflows were deduced from reservoir levels during several storms; the median unit hydrograph from five storms was accepted as the design unit hydrograph. The percentage runoff model of the *Flood Studies Report* (NERC, 1975) was adapted to local observations, with a standard percentage runoff of 10% increasing to 24% for the 20-year event and 39% for the 500-year flood. The 500-year peak was estimated as over three times the 20-year peak before reservoir routing.

ESTIMATION OF MAXIMUM FLOOD

For major projects where no significant risk of overtopping could be accepted, the estimation of the maximum flood is normally specified. The "probable maximum flood" may be described as the physical upper limit of flood discharge at a site. This was the recommendation of the Institution of Civil Engineers (1978) for major dams in the UK.

There must be an element of judgement in the various stages of estimating a maximum flood. The estimation of the maximum storm and its time profile may be straightforward if a comprehensive meteorological study has been carried out, but this is the exception in most dam design investigations in developing countries. However, with the increasing acceptance of the concept of the maximum flood, the basic meteorological analysis has often been carried out on a regional basis. For example, the review of flood design for two projects in India in the 1980s involved individual estimates of the storm magnitude, while a similar review in the 1990s could take advantage of a countrywide analysis of maximum storms. The conversion of storm estimate to runoff hydrograph is more clearly linked to local conditions and depends crucially on storm losses; the estimation of these losses often involves extrapolation from conditions already sampled to exceptional conditions. Two examples illustrate some of the problems normally encountered. One of these involved a dam on the Yarmouk, a tributary of the Jordan draining a semiarid basin in Jordan and Syria; the other was the Victoria dam on the Mahaweli Ganga in central Sri Lanka.

Jordan

In the case of the Jordan site, the maximum rainfall was estimated by transposing a historic storm, which covered the east bank of the Jordan valley, to the basin of the Yarmouk. This storm was then maximised by comparing the available moisture during the historic storm and the maximised moisture content and efficiency derived by the UK Meteorological Office during a regional study of potential floods in the Jordan valley. This led to an estimated 24-hour maximum storm of 245 mm, which exceeded the historic storm by a factor of 4.5. However, during the review of the study the estimate was reduced to take account of seasonal variations in the maximum moisture available. A number of observed storms were examined to derive depth–duration relationships, from which an envelope curve was used to give incremental depths for the design storm. These rainfall increments were arranged in an order which led to a critical flood-producing sequence.

Estimates of the potential basin losses were based on losses appropriate for comparable soil types in the USA, and the rainfall excess was distributed according to a 4-hour unit hydrograph developed for the dam site. The initial peak flow estimate at the site, before reservoir attenuation, was 8800 $m^3\ s^{-1}$, which compared with an estimated peak flow of 50-year return period of 800 $m^3\ s^{-1}$, based on statistical analysis of 15 years of flood peaks. The estimate of over ten times the 50-year estimate reflected the contrast between the normal storm in these semiarid conditions, and the potential precipitation in an extreme storm, but also included an element of conservative design which was corrected at the review stage.

Sri Lanka

The basin of the Mahaweli Ganga at the Victoria dam site below Kandy (Piper *et al.*, 1994) was, by contrast, mountainous and humid (Plate 10.3), with rainfall occurring in each month. The average basin rainfall at Gurudeniya, the nearest river gauge upstream, was 2800 mm. As peak flow records were limited in duration and based on extrapolation of rating curves, the use of flood frequency methods for spillway flood estimation would not be precise. On the other hand, the river basin had been developed for tea plantation and a dense network of rainfall gauges had been in operation since about 1880. The most reliable approach was therefore the estimation of the maximum

Plate 10.3 *Tributary of Mahaweli Ganga between Kandy and Nuwara Eliya, Sri Lanka.*

precipitation and its conversion to a flood hydrograph using the unit hydrograph method. An annual isohyetal map (Fig. 8.1) was compiled from the records of over 100 stations maintained by the Meteorological Department. Another map was compiled showing the date and amount of the maximum daily rainfall recorded at each station; this statistic was conveniently listed in yearbooks. Because of the large number of long-term gauges, this gave a useful indication of the likely magnitude of the maximum storm over the basin. It was observed that many of the stations recorded their maximum storm around the same date: 14 August 1947. This simplified the subsequent analysis. A number of recording gauges had been installed more recently, and the data had been used in a rainfall depth–duration–frequency study for the whole of Sri Lanka (Baghirathan & Shaw, 1978).

An earlier unpublished study (Sumanasekere, 1977), based on personal research by a member of the staff of the Meteorological Department, provided much of the material for an analysis of moisture maximisation. He had examined the 30 heaviest storms and isohyetal maps had been drawn for each storm from daily data; this confirmed that the storm of August 1947 was the highest on record. It also showed that all the large storms had a similar shape, orientation and centre, presumably because of the mountainous topography. It was not considered appropriate to transpose the centre or the orientation of historic storms to maximise the design storm. As the August 1947 storm was the highest storm recorded at 21 out of 37 stations, had covered much of the upper basin, and had led to the largest flood on the upper Mahaweli since records began in 1888, it was decided to use a simple method of maximisation of this storm to estimate the maximum flood.

An observed storm may be maximised for orientation, wind speed, moisture content and storm efficiency, but as many of the large historic storms covered the

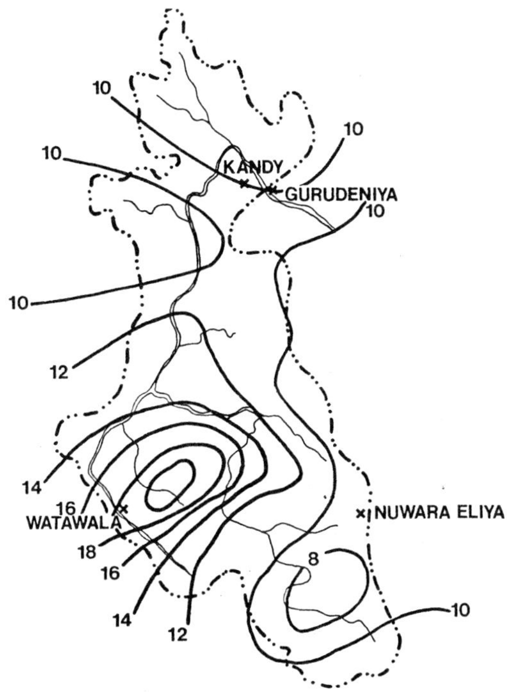

Fig. 10.11 Isohyetal map of daily storm of August 1947 above Gurudeniya. Note: the units are inches, 1 inch = 2.54 cm.

same, extensive area of the upper basin, it was assumed that the storm efficiency had approached the maximum possible. The 1947 storm (Fig. 10.11) was maximized for moisture content alone, using data from the climate stations at Kandy and Nuwara Eliya. Comparisons of the maximum dewpoints ever recorded at these sites with those measured during the 1947 storm were used to estimate the ratio of the maximum precipitable water to that during the actual storm. The design storm over the project basin was estimated by multiplying the isohyetal map by this ratio, which was relatively modest at about 1.25. Statistical analysis suggested that this storm lay between estimates of the 1000-year and maximum storms. The duration and time profile of the design storm were based on the profile recorded in 1947.

The conversion of the storm profile to the corresponding flood hydrograph followed the principles of the *Flood Studies Report* (NERC, 1975), namely the derivation of a unit hydrograph from historic storm and flood records, the conservative estimation of basin losses and the convolution of rainfall excess with unit hydrograph ordinates. Only two events were considered suitable for unit hydrograph derivation, but tests of the derived unit hydrograph were judged satisfactory. The percentage of storm rainfall contributing to flood runoff is critical, but is related to basin characteristics, antecedent basin wetness and storm size. On this basis, the percentage runoff observed in 1947 of 56% was increased to 70% for the design case. The resulting design flood hydrograph is illustrated in Fig. 10.12, with a peak flow of 9500 $m^3\ s^{-1}$, and compared with an estimated 1000-year flood, based on a storm derived from statistical analysis, with a peak flow of 6400 $m^3\ s^{-1}$. This flow compares with a mean annual flood of 1400 $m^3\ s^{-1}$ at Gurudeniya upstream.

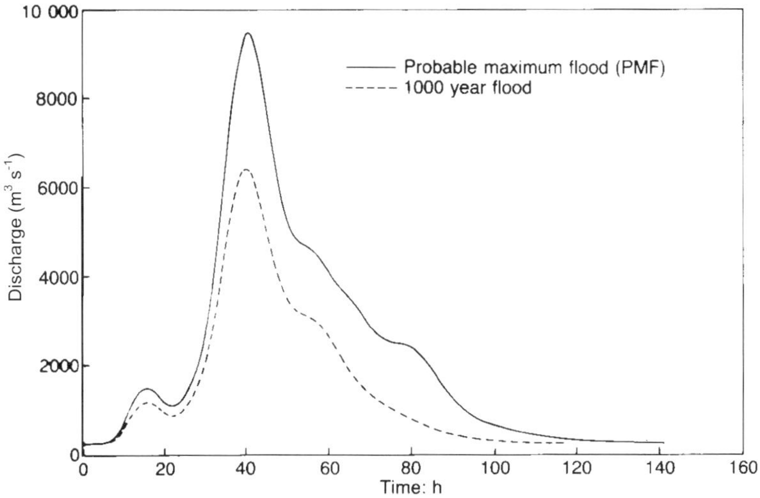

Fig. 10.12 *Design flood hydrograph: Victoria dam, Sri Lanka (from Piper et al., 1994).*

This result contrasts to the maximum flood in Jordan, where the design flood was a much higher multiple of the mean annual flood. It illustrates the point that the maximum flood is more likely to be approached in a historic record in a humid than an arid climate. Meigh *et al.* (1999) have shown that although mean annual floods are much lower in areas of low rainfall, the growth curves are much steeper than in humid climates where the mean annual flood is much higher. Indeed, in the UK it has been shown (Farquharson *et al.*, 1975) that the maximum floods vary less from west to east in the direction of decreasing precipitation than mean annual floods or floods of moderate return period.

CONCLUSION

There are several different methods of estimating design floods, but the choice depends largely on the records which are available, and the return period required for the particular project. Statistical methods of analysis are greatly strengthened by increasing the effective length of record, either by grouping flood records from various sources or by incorporating historical information. The influence of climate on flood frequency relations is very significant. Physical methods may be based on storm analysis and evidence of the conversion of storm to flood events.

11 SEDIMENTATION AND ENVIRONMENTAL ISSUES

Some topics which are important to water resources studies do not fit easily into the sequence of processes of the hydrological cycle. These topics include the estimation of sediment transport in rivers, including both suspended sediment and bed load, and the assessment of the impact of water resources projects on the local environment and economy. A few examples are included, based on problems which have been encountered. The treatment is not as complete as the topics deserve; their inclusion is intended to draw attention to the need for such studies.

SEDIMENT TRANSPORT

The assessment of sedimentation is an important part of reservoir design, especially in the more arid parts of the world where reservoirs are vital. Reservoirs in such areas are particularly vulnerable to sedimentation, because the few intense flows are likely to be heavily sediment-laden. This can severely limit the useful life of reservoirs unless the sediment load is estimated with care and appropriate reservoir operation is designed.

However, sediment transport is not easy to measure or to estimate. Sediment load is linked to river discharge but is highly sensitive to variations in flow. Further, the relation between river flow and sediment load is not constant, as it depends on the supply of material eroded from the basin and is greater at the beginning of the wet season and after heavy storms. Sediment load is also sensitive to changes in basin vegetation, due for example to deforestation.

The load consists of two main components, suspended sediment and bed load. Suspended sediment can be sampled at river gauging stations, though it varies with depth; bed load is more difficult to measure, as it may only occur during high flows when it is not easy to sample. The problems are best discussed with specific examples.

Tana basin, Kenya

During investigation of the Tana basin in 1958 (Chapter 3; Sutcliffe & Rangeley, 1960a), it was necessary to estimate the sediment load within the basin as this would affect the economic life of any reservoir to be constructed. This was made possible by the fact that about 100 suspended sediment measurements had been made at each of a number of gauging stations in a recent research programme. These measurements were based on a depth-integrating sampler suspended from a cableway; the sampler was designed to be lowered from the water surface to the bed and raised at a constant rate so that the sediment intake would sample the total load over the profile. The measurement was repeated at three points on the cross-section, and the mean analysed

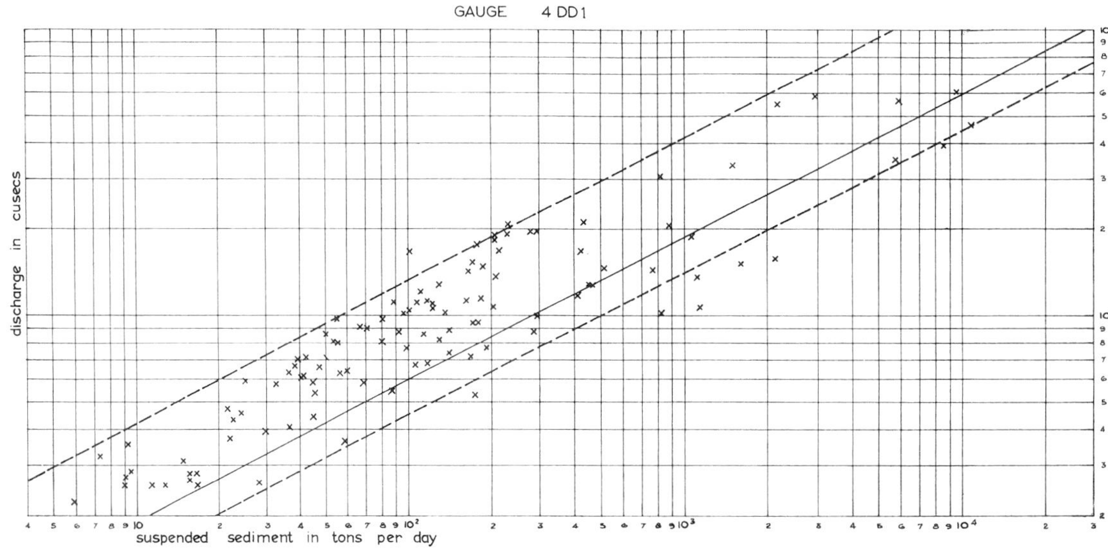

***Fig. 11.1** Thiba above Kamburu: sediment rating curve. Note that the units on the axes are log(cusecs) and log(tons per day).*

sediment concentration was calculated. This concentration could be multiplied by the river discharge to give an estimated suspended load in tons per day.

At ten stations sufficient results were available for sediment load to be compared with discharge. Although no account was taken of the season or whether the river was rising or falling, there was a clear correlation between sediment and discharge at each station. The relation appears as a linear band when plotted logarithmically; an example is illustrated in Fig. 11.1. At most sites the scatter was considerable and the range of loads for a given discharge was found to vary by an order of magnitude. Because of this scatter, and the effect of the logarithmic scale, a graphical approach was preferred to statistical correlation. Two parallel lines were drawn to envelop all but a few outlying points, and a third line was drawn to give the arithmetic mean of the two envelope lines, taking account of the logarithmic scale.

The estimation of total load involves the conversion of river flow to sediment load by means of the sediment rating curve. Because the sediment concentration is highly sensitive to flow, there had to be a compromise between conversion on a daily basis, which would have involved a considerable amount of calculation before the availability of computers, and the calculation of sediment loads on a monthly basis, which would greatly reduce the peak flows and concentrations. In this case the suspended sediment loads at six sites were calculated using 5-day or 10-day periods and mean flows were converted to sediment loads in tons per day. The total loads were deduced for each year. The annual sediment loads for the Tana at Kamburu, the most complete station, are compared with the annual flows in Fig. 11.2; it is evident that the transport of suspended sediment is extremely sensitive to high flows and that a large proportion of the long-term load is carried in years of high river flow. The annual average sediment load was calculated for the five tributaries and the main river at Kamburu; it was found that the total tributary load was higher than the downstream estimate, which may have

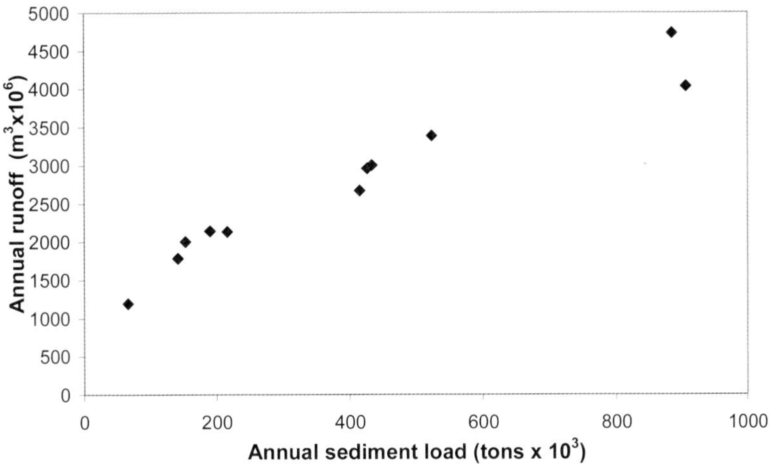

Fig. 11.2 *Tana at Kamburu: annual sediment and runoff:*

been due to the short period of sampling, the rating curves derived, or deposition above the lowest site. No information was available on bed load, so the results were confined to suspended sediment. The results confirmed the visual impression that the rivers flowing from Mount Kenya carried less sediment than the rivers draining the Aberdare range, while the tributaries whose basins had at the time a higher proportion of cultivated land carried more sediment than those which were primarily used for grazing.

Some 20 years later this same estimate was used by others as the basis for the design of a proposed reservoir, as no subsequent analysis was apparently available. This was criticised on the evidence of more recent measurements, which suggested that the sediment load of the Tana tributaries had increased dramatically over the 20-year period, with greatly increased intensity of agriculture and changed land use in the headwaters following population pressure. This illustrates the sensitivity of erosion and sediment load to changes of land use, which had been investigated on an experimental basin scale upstream by the East African Agriculture and Forestry Research Organization (EAAFRO) and later investigators (Pereira, 1962; Blackie *et al.*, 1979).

Bed load estimation

In a field investigation on the River Tongariro in New Zealand in 1959 (Sutcliffe & Rangeley, 1960a), similar suspended sediment samplers were constructed and used to monitor sediment loads. Additionally, it was evident from the channel material that bed load movement was likely to be a significant component of the total transport of this mountain river and its tributaries. A bed load sampler was tested on the main Tongariro River, but its use was impractical during the periodic spates in which the load was concentrated. The bed load was therefore estimated instead at a water supply intake site below the main skiing area on Mount Ruapehu, where the intake basin had to be cleared of bed load, including boulders, at regular intervals. The average annual volume removed gave an estimate of the tributary bed load.

In a review of solid discharge estimates for the Tarbela dam project on the main Indus in Pakistan, it was noted that the bed load estimation had previously been based on a nominal percentage of the suspended load. This was tested by using semi-empirical formulae relating bed load movement to channel gradient, river velocity, and the typical size of bed load material; the last was measured during a field visit. This illustrates a recurrent theme of this book, that there is no single correct way to estimate hydrological processes; methods of investigation depend largely on the evidence which is already available or which can be assembled during the time available.

Reservoir operation to minimize sedimentation

The Blue Nile provides an example (Sutcliffe & Widgery, 1997) of a river where the sediment load is high, particularly compared with the size of the reservoirs which can be constructed within the Sudan to store water required for local irrigation and for hydroelectric power production. The sediment load of the Blue Nile above Roseires has been estimated as 140 million tonnes per year (El Moushid *et al.*, 1997). The distribution of sediment during the flood season is important. The concentration of suspended sediment reaches a peak in late July or early August before the peak discharge is reached, and decreases until the end of the flood; an example is illustrated in Fig. 11.3. The storage available at Roseires reservoir and at Sennar downstream is fairly limited and it is important to minimise the sediment deposition by careful operation of the reservoirs to pass the flow during the flood period. In 1995, after nearly 30 years of operation, the capacity of Roseires reservoir had been reduced by 0.9 km^3 to an estimated 2.4 km^3, compared with an average annual flow of 49 km^3. The strategy has been to delay the filling of the reservoirs until the peak of the flood has passed through the reservoirs, as the sediment concentration is greatest on the rising limb of the hydrograph. Flow forecasting from rainfall could be useful in postponing the filling as long as possible while minimising the risk of not filling the reservoir.

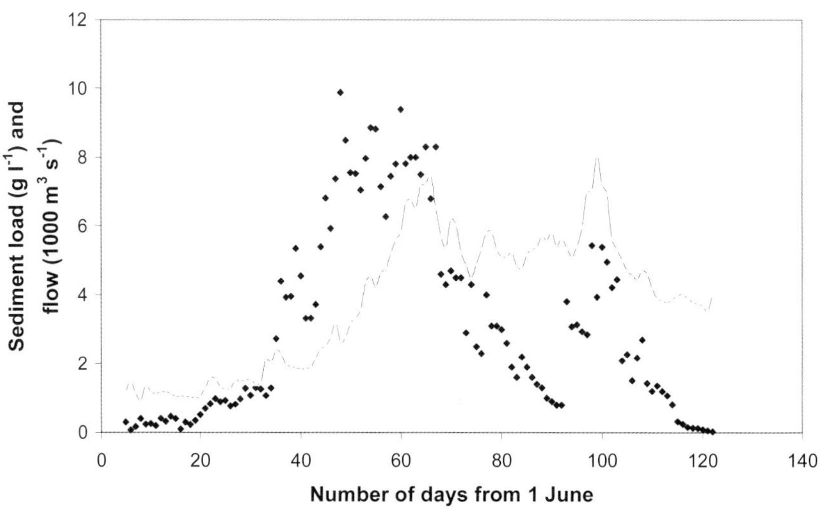

Fig. 11.3 *Blue Nile at Roseires: sediment concentration; ♦ sediment, – – flow.*

ENVIRONMENTAL ASPECTS

The following examples of environmental studies involve the effect of different types of projects on African wetlands. In one case, the project was the Jonglei Canal, whose purpose was the reduction of evaporation losses in the large Sudd wetlands in the southern Sudan in order to increase the flows of the White Nile downstream. In this case the emphasis of the study was the effect of the project on the natural flooding of the Sudd wetland, and the result which this would have on the vegetation of the flood plain. This study thus included the mechanism by which the seasonal flooding controlled the natural vegetation, and the change of vegetation which would result from the change of regime caused by the project. In another case, potential reservoir storage in Lake Albert would cause flooding of areas used for cultivation and also areas of wetland above and below the lake; the level to which flooding would reach was the subject of investigation. The proposed use of the River Kagera for hydroelectric power production would also result in flooding upstream of the reservoir and a change in regime through the wetlands on the lower Kagera; the ecological control of the wetland vegetation was similar to that deduced in the Sudd. In the lower Senegal valley, where natural flooding produced vegetation similar to that of the Sudd, areas of seasonal flooding were also used for cultivation when the flood had receded; a reservoir built upstream for power production and irrigation was operated initially to simulate the natural flooding but with greater reliability.

Study of Jonglei Canal and Nile wetlands

The assessment of the environmental impact of projects has become an essential part of most investigations in recent years. In 1950, when the author joined the Jonglei Investigation Team, the concept was relatively new. However, the team had been set up by the Sudan Government to investigate the potential effects of the Egyptian Government's proposals to increase the contributions of the White Nile by bypassing the Sudd. Its report (Jonglei Investigation Team, 1954) has been described as the first environmental assessment in the developing world. The scale of the effort resulting in this report may be illustrated by the fact that a team of ten expatriate professional staff, with local support including a Nile steamer (Plate 11.1), spent up to four years on the study. The following summary of the hydrological aspects of this and later studies is divided into accounts of the physical background, the effects of successive projects for the Jonglei Canal, and the prospects for mitigating these effects.

Because it had been noted early in the 20th century (Garstin, 1904) that the outflow from the Sudd was only about half the inflow of the major channel, it was suggested that a diversion channel could reduce the evaporation losses. For this reason the flows through the Sudd have been measured and the hydrology investigated over much of the 20th century. After the economic importance of the dry season grazing provided by seasonal inundation had been noted, complementary ecological studies of the area were combined with hydrological work in, for example, the research of the Jonglei Investigation Team.

The physical background

The basin of the Bahr el Jebel at Mongalla, at the upper end of the Sudd, includes Lakes Victoria, Kyoga and Albert (Fig. 11.4). The inflow to the Sudd combines a lake-

Plate 11.1 *Papyrus swamp on the Bahr el Jebel.*

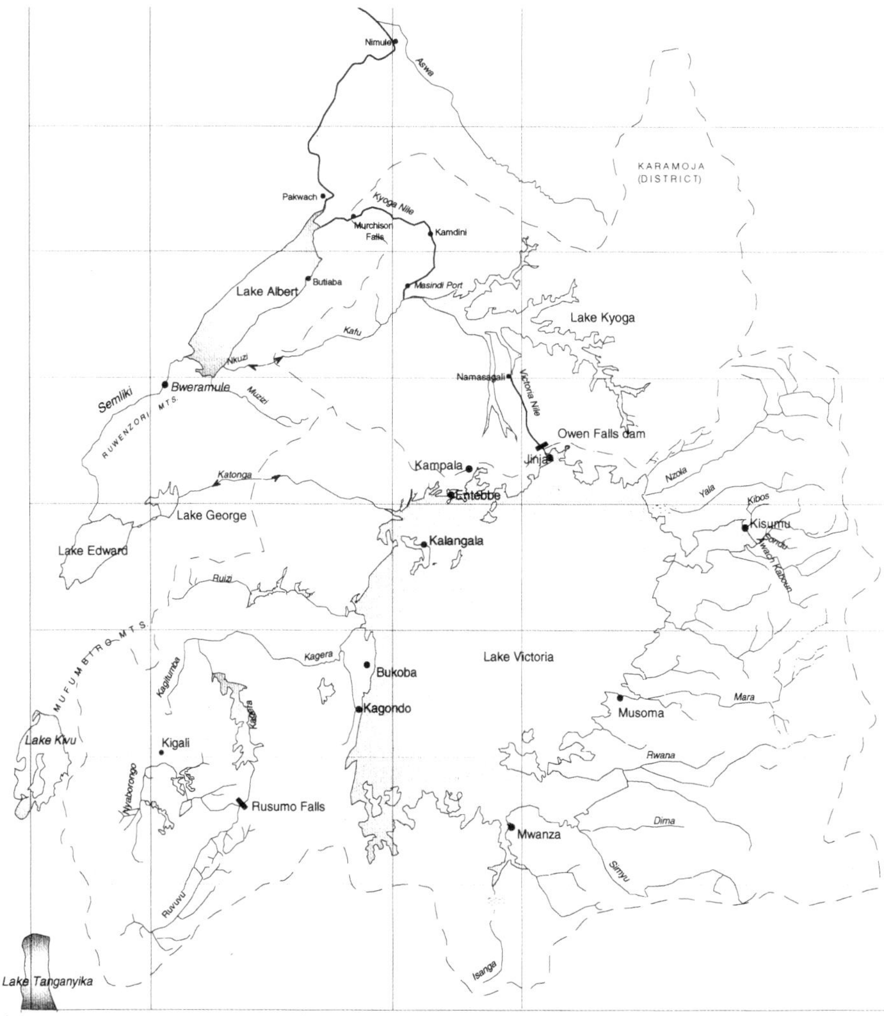

Fig. 11.4 *Map of the upper Nile basin (from Sutcliffe & Parks, 1999).*

fed component with the highly seasonal torrent inflows below Lake Albert. The outflow from Lake Victoria was controlled by lake level at the Ripon Falls until the Owen Falls dam was constructed in 1951–1953; since then the dam has been operated to mimic the natural outflow. The lake level (Fig. 8.26) was fairly stable between 1896 and 1961, but rose 2.5 m in 1961–1964 and the outflow doubled (Fig. 8.27). The average rainfall over the lake is very similar in amount to the lake evaporation. Thus an increase in lake rainfall, which occurred after 1961 with the associated rise in tributary inflow, had an exaggerated effect on lake level and outflow. After the dramatic rise in 1961–1964, this increase in rainfall continued to affect lake level and outflow though the lake level has fallen. There is evidence, from eyewitness accounts and from downstream flows, that high levels also occurred around 1878 and 1895, though the reversion to lower levels was faster than in recent years (Tate *et al.*, 2001).

The net effect of Lakes Kyoga and Albert has been to enhance the variations of the outflows from Lake Victoria. Before 1961 there was a small decrease between inflows and outflows through Lake Kyoga; since 1961 there has been higher outflow than inflow. A similar change occurred in the regime of Lake Albert; after 1964 there was a net contribution from the lake basin. The torrents below Lake Albert reflect their single rainfall season and provide the highly seasonal element of inflow which causes the seasonal flooding of the Sudd.

The water balance of the Sudd

Within the Sudd, the Bahr el Jebel flood plain (Fig. 11.5) is incised several metres below the surrounding woodland. This valley widens from about 5 km near Mongalla to 15 km north of Bor, some 120 km downstream, and further north the woodland and the incision disappear. The alluvial river channels are built up above the flood plain; these channels are inadequate for all but the lowest inflows. Higher flows spill through or over the alluvial banks and flow down the flood plain and partly return to the river channel downstream. As the inflow to the Sudd increases seasonally or after upstream lake rises, wider areas are inundated and evaporation from these areas increases. As a result the channel flows decrease steadily through the Sudd and the eventual outflows (Fig. 11.6) lose their seasonal variation and are only about half the inflows.

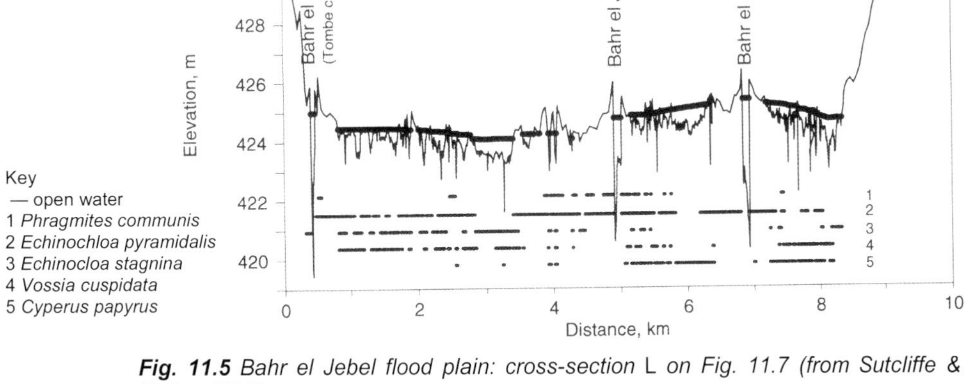

Fig. 11.5 *Bahr el Jebel flood plain: cross-section* L *on Fig. 11.7 (from Sutcliffe & Parks, 1999).*

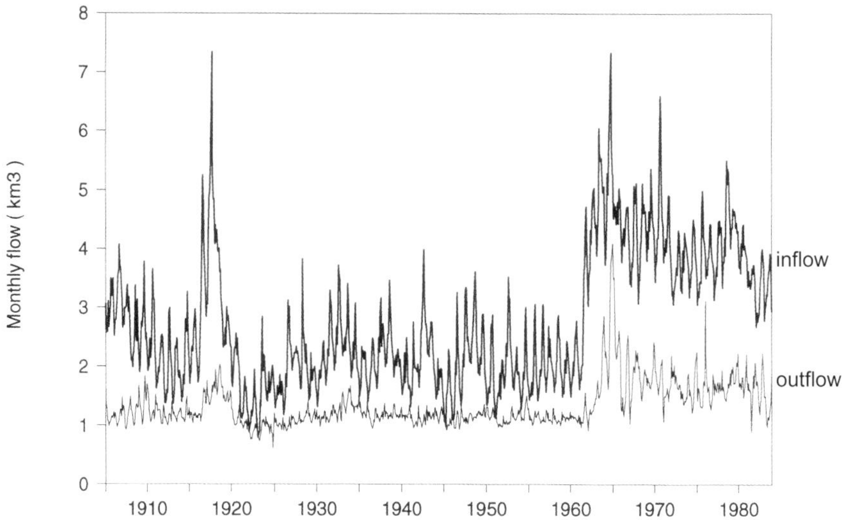

Fig. 11.6 *Sudd inflows and outflows (from Sutcliffe & Parks, 1999).*

From measurements of inflows and outflows for either a sample reach or the whole of the Sudd, together with estimates of rainfall and evaporation over the flooded areas, a water balance can be used to estimate the areas inundated over the period of records. This also requires information on the relation between areas and volumes of flooding. In 1951–1952, the Jonglei Investigation Team carried out a survey of the southern Sudd between Juba and Bor, which included 16 cross-sections of the whole valley (Fig. 11.7), 15 sections of the Aliab valley to the west of the main river, and 12 sections of the Mongalla basin to the east; these surveys of sample areas included longitudinal surveys of the alluvial banks of the river, transects of the vegetation and soils along the cross-sections, and also studies of the use of the flood plain for grazing.

Hydrological control of vegetation

A water balance of the Mongalla basin, between Mongalla and Gemmeiza, showed (Sutcliffe, 1957, 1974) that the volume of water in transit through the basin was directly related to the inflow, and therefore the volume of seasonal flooding could be related to the flow records at Mongalla. From the recorded vegetation along each cross-section (Fig. 11.5), the distribution of species with elevation could be deduced along each cross-section and down the flood plain. It was evident (Fig. 11.8) that there was a distinct elevation boundary between the deep-flooded flood-plain species (*Echinochloa stagnina*, *Vossia* and *Cyperus papyrus*) and the shallow-flooded species (*Echinochloa pyramidalis*, *Oryza* and *Phragmites*). This vegetation boundary was traced along the river profile and shown to be parallel to the high river level; it was deduced that this boundary was controlled by the maximum depth of flooding. From the statistics of flooding volume it was estimated that this maximum depth of flooding was 1.30 m. Because papyrus was limited to the lower end of each basin, and the elevation range of flooding increases from the lower to the upper end of the basin, it was also deduced that the presence of papyrus, anchored by fragile rhizomes, is controlled by the range of flooding; the limiting vertical range was 1.50 m. These conclusions on how the hydrological conditions

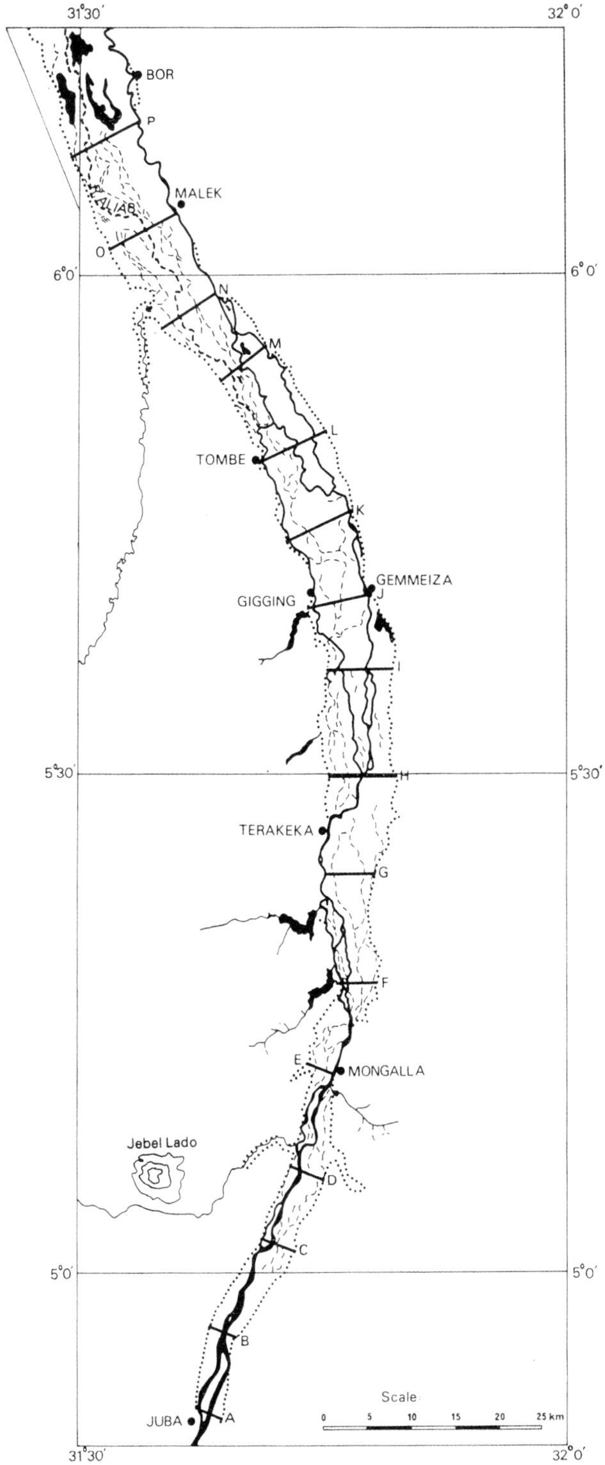

Fig. 11.7 *Map of the Bahr el Jebel flood plain between Juba and Bor (from Sutcliffe & Parks, 1999).*

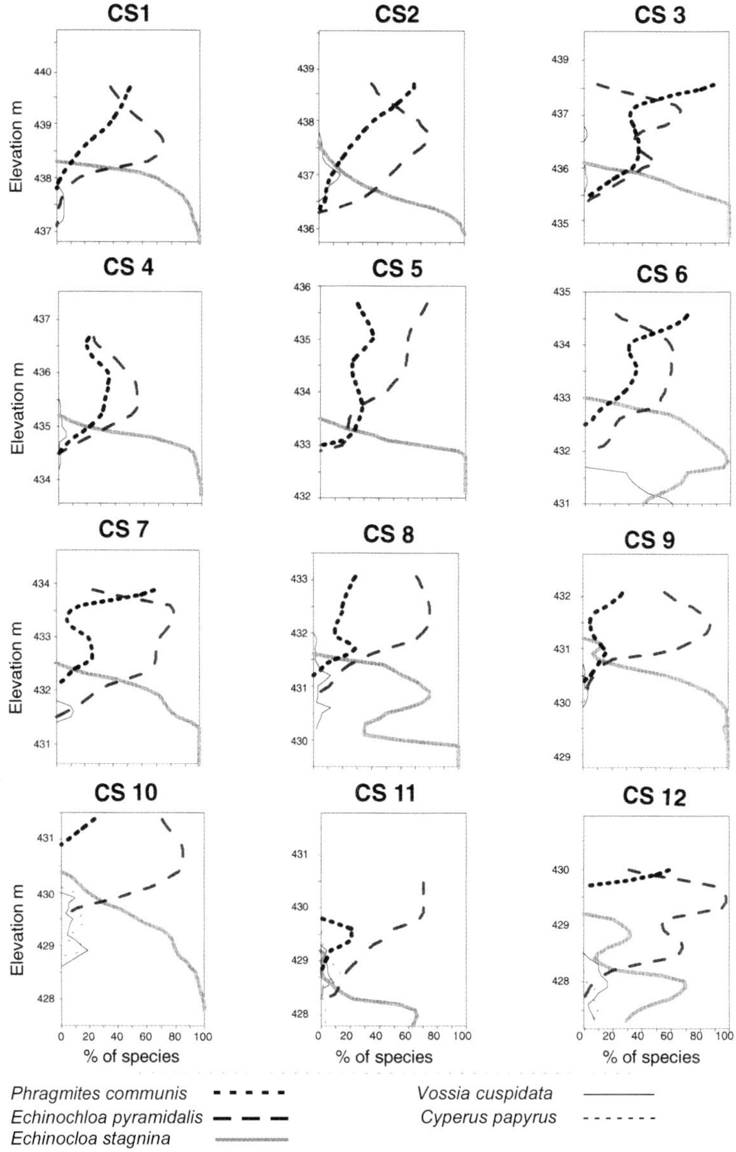

Fig. 11.8 Vegetation distribution in the Mongalla basin: cross-sections in order from south to north (after Sutcliffe & Parks, 1996).

control the vegetation (Sutcliffe & Parks, 1996) were consistent with the areal distribution of species, and also with observations elsewhere in the Sudd of the flood depths recorded where species were found.

The distribution of these vegetation species on the flood plain was important since the pastoral economy relied on the dry season grazing provided by the *Echinochloa* species and *Oryza* (Plate 11.2). At the wet end of the distribution, the papyrus swamp provides no grazing, while the *Phragmites* at the dry end is unpalatable. The local economy was based on the annual migration to the flood plain during the dry season,

Plate 11.2 *Cattle grazing on the Bahr el Jebel flood plain.*

and on rainfed vegetation during the rest of the year. The maintenance of the annual inflow regime, resulting in the annual inundation of the flood plain through the torrent flows, and its uncovering during the course of the dry season, was essential to the pastoral economy; alternative agriculture on higher ground was difficult because of the frequency of either drought or excessive flooding.

Impact of the Jonglei Canal

Because of the Bahr el Jebel losses in the Sudd, there have been several proposals since 1904 to reduce the losses of flow downstream by carrying part of the flow past the Sudd in a channel. Some proposals included regulating the inflows by storage upstream in Lake Albert or Lake Victoria. Others were confined to banking the Bahr el Jebel, but most recent plans have been based on a canal, known by its offtake site as the Jonglei Canal.

Detailed proposals for the Equatorial Nile Project were published by the Egyptian Government in 1946 (Hurst *et al.*, 1946). These proposals included storage in Lake Albert controlled by a dam, either immediately below the lake at Mutir, or at Nimule some 200 km downstream, together with a regulator at Jinja on Lake Victoria. This would have allowed the control of Lake Albert outflows, including the virtual storage of the torrent flows, and their release at a time when they were required for irrigation in Egypt. To reduce transmission losses in the Sudd, flows of 55 $m^3 \times 10^6$ day^{-1} would be diverted through the Jonglei Canal and the balance passed down the natural river. It was planned that the storage could be used at a later stage to increase flows down the main Nile during years of deficit on the Blue Nile, with the Jonglei Canal expanded to pass these flows. The Jonglei Investigation Team studied the impact of these proposals and concluded that the associated reversal of seasonal flows above the canal intake, the maintenance of low flows throughout the year in the natural river below the intake, and of high flows below the canal outfall, were unacceptable. A Revised Operation was proposed which would have greatly reduced the adverse effects of the project.

After 1956 attention turned to the Aswan High Dam, with overyear storage at one site. This removed the importance of the flow timing for irrigation demand in Egypt, which had led to the planned reversal of flows below Lake Albert. Revised plans for the Jonglei Canal were agreed between Egypt and Sudan, aimed at increasing annual downstream flows by 4 km^3 by reducing flows through the natural river. This did not involve the reversal of flows, and the environmental issue was the reduction of the extent of the inundated areas, particularly the important seasonally flooded area.

The construction of the revised Jonglei Canal (Fig. 11.9), which had little similarity to the earlier project, began in 1978 but has been interrupted since 1983 to the present. Its potential effects were studied by Mefit-Babtie, with a hydrological assessment by the Institute of Hydrology (Sutcliffe & Parks, 1982, 1987; Howell *et al.*,

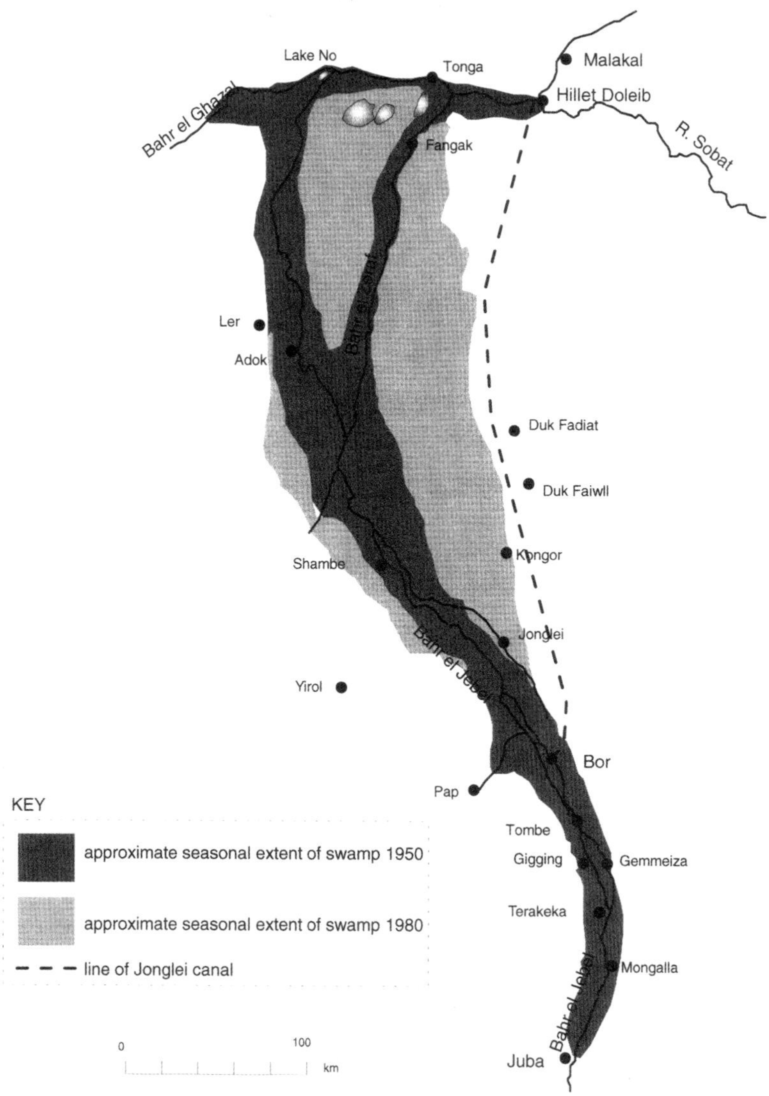

Fig. 11.9 *Map of the Sudd area (after Howell et al., 1988).*

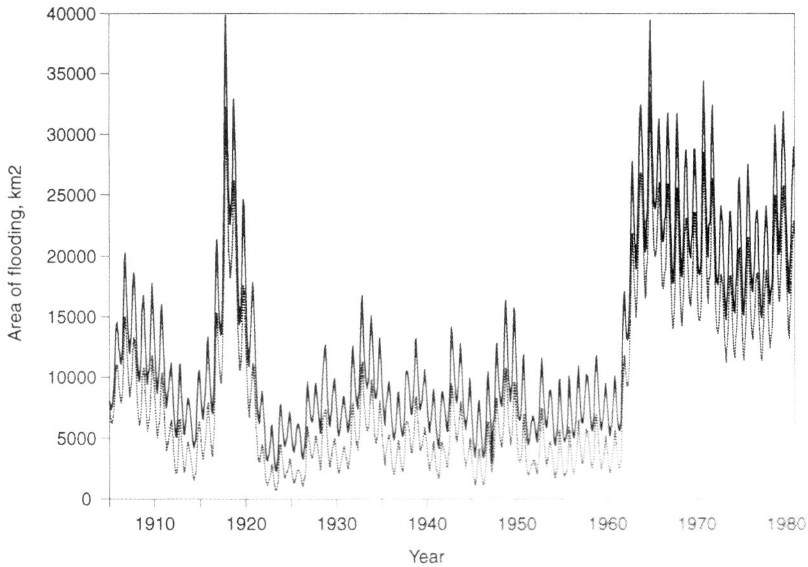

Fig. 11.10 *Effects of the Jonglei Canal: estimated areas of flooding below Bor, with and without the Jonglei Canal (20 m^3 x 10^6 day^{-1}) (from Sutcliffe & Parks, 1999).*

1988). The hydrological study was based on the water balance of the whole Sudd, using measured inflows and outflows for the period 1905–1980, rainfall measurements around the Sudd and evaporation from flooded areas using Penman estimates of open water evaporation. The model included soil moisture recharge of newly flooded ground, and used a simple linear relation between area and volume of flooding, which implies a constant depth as flooding spreads.

This model was run over the historic period to estimate the areas of flooding (Fig. 11.10), and reproduced reasonably well both the seasonal fluctuations and the changes after the rise in Lake Victoria. Using a relation between measured inflows and outflows and applying the same relation to river flows after construction of the canal, the model was used to predict the areas of flooding (Fig. 11.10) which would have occurred with the Jonglei Canal operating with a constant discharge of 20 m^3 × 10^6 day^{-1}. This suggested that the areas of seasonal flooding would have decreased by some 26% in the lower flow period of 1905–1961, and by 17% in 1961–1980; the permanent swamp would have decreased by 46% and 21% respectively, in the same periods. It was important to note that the timing of the seasonal inundation and uncovering would not have been affected.

Mitigation of impact

The hydrological modelling also suggested that the adverse impact could be reduced by varying the canal flows seasonally. Trials suggested that altering the canal flow to 25 m^3 × 10^6 day^{-1} from November to April and 15 m^3 × 10^6 day^{-1} from May to October would lead to a decrease in seasonal flooding of only 12% in the early period and 10% in the later period. By varying seasonally the flow down the canal, the natural river flows would be increased during the wet season and decreased during the dry season; this would enhance the natural fluctuations and thus increase the areas of seasonal flooding at the expense of permanent swamp. Equally importantly, it would increase the range of depth of flooding, which would encourage the valuable grazing grasses rather than papyrus.

Observed changes, following the rise in Lake Victoria, in the hydrological regime of the Sudd and the resulting vegetation support the deductions about the links between flooding and vegetation. During the 1982 hydrological study, the opportunity was taken to observe the Aliab valley from the air at low level along the survey cross-sections, armed with vegetation and other maps of the 1951 survey. It was noted (Fig. 11.11) that *Vossia* and papyrus had spread over areas previously dominated with grazing grasses. Because the seasonal torrent flow volumes had not increased after 1961, while the area of flooding had doubled, the range of depth of flooding would have decreased to favour papyrus. The *Vossia* would have been encouraged by the higher flow velocities. In the drier Mongalla basin, the increased flooding had encouraged the spread of grazing grasses, and cattle were observed grazing areas which had previously been dominated by *Phragmites*.

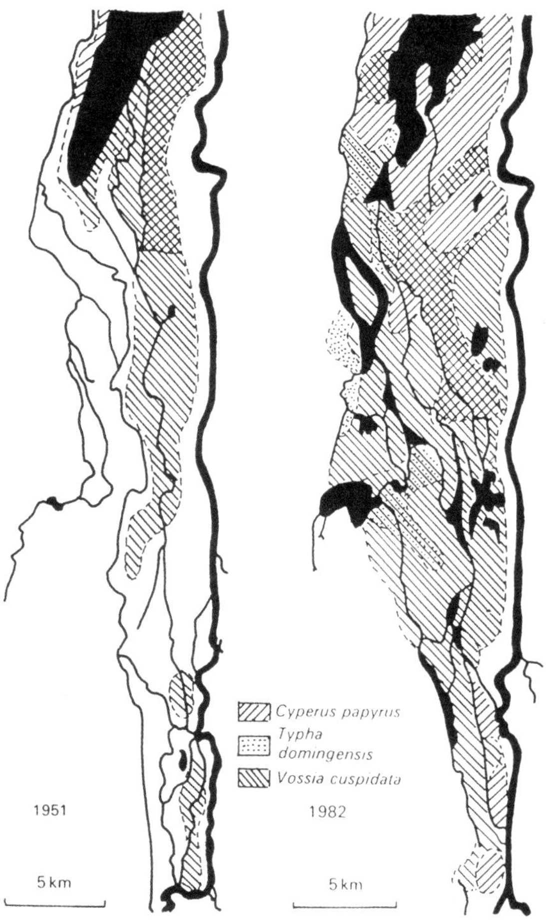

Fig. 11.11 *Vegetation changes in the Aliab valley between 1951 and 1982 (from Howell et al., 1988).*

Thus the potential effects of the Jonglei Canal project which was started in 1978 may be predicted in general terms. Above the canal offtake, the scheme would have no effect in the absence of upstream storage. Between the canal offtake and outlet near the Sobat confluence, the project would not affect the seasonal pattern of flooding but would reduce the areas of seasonal flooding; the effect would be less than on the less valuable permanent swamp. Indeed, it would be possible to reduce these adverse effects by suitable adjustment of the canal inflows. Below the canal outfall, the flows down the White Nile would be increased by the reduced losses in the Sudd; the seasonal distribution of flows down the White Nile is dominated by the Sobat inflows.

Local effects of storage in Lake Albert

After the report of the Jonglei Investigation Team (1954) had been published, negotiations between Egypt, Sudan and Uganda were due to take place over the use of Lake Albert as a reservoir to supplement the Jonglei Canal. For this purpose the Uganda Government needed information on the area which would be flooded by a dam at Mutir, below the exit from Lake Albert. This required a topographic survey of the area around the lake, supplemented by a study of the soils and crops and an estimate of the population involved. In order to carry out a rapid survey at this stage of negotiations, a simple instrument was used to measure cross-sections away from the lake; this consisted of two level staves with a transparent plastic tube inserted vertically and connecting the staves and filled with water. The water level in the tube was read against each level staff and the difference in level deduced. This procedure was continued along the section until the required elevation above the lake surface was reached, and thus a section could be plotted without the need to clear sightlines which form the main expense of conventional surveys.

A contour map was compiled by interpolation and the soils and crops were examined and classified within the limits of the land liable to inundation. The extent of wetland vegetation around the lake was also noted. In the event the negotiations were overtaken by the decision to investigate the Aswan High Dam, where Nile flows are now stored at a location with significantly higher evaporation. However, the Lake Albert dam is again being discussed. It is probable that the downstream effects along the Bahr el Jebel flood plain of operating such a dam could be more significant than the local inundation.

The Kagera flood plain

Proposals have more recently been made to develop hydroelectric power near the Rusumo Falls on the Kagera above Lake Victoria. There are lakes and wetlands above and below the falls, and the flood-plain vegetation species are similar to those on the Bahr el Jebel. It follows that the vegetation upstream and downstream would be affected by reservoir storage and in particular by changes in the depth and range of flooding. The general effects on vegetation could be predicted on the basis of the ecological controls deduced from the Sudd.

Control of Senegal flows to maintain flooding

An example of the control of a reservoir to improve the natural environment is given by the River Senegal. This river (see also Chapter 9), whose main sources are in the

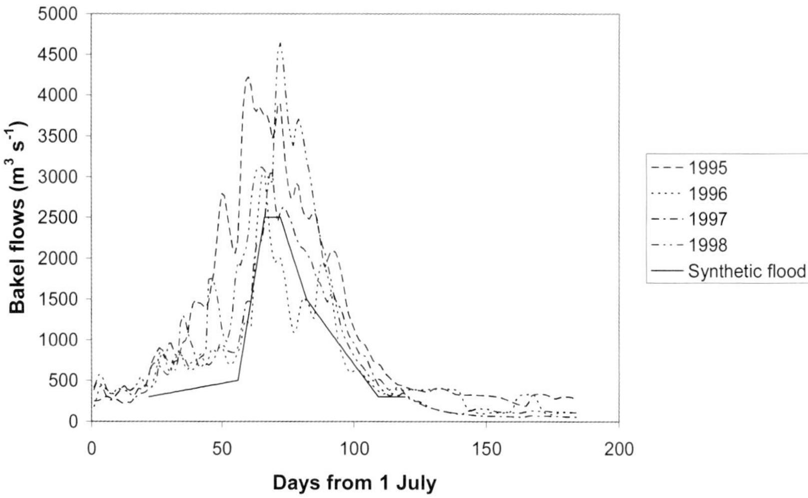

Fig. 11.12 *Senegal at Bakel: flows, 1995–1998.*

mountains of the Fouta Djallon in Guinea, in its natural state inundated an area of wetland in its lower reaches which averaged about 5500 km^2, but which varied from a very small area in a dry year to about 8000 km^2 in a wet year. The seasonal flooding provided the basis of the local economy, including dry season grazing for about a million cattle, recession agriculture on the area of flood plain where soil moisture was recharged by the annual flooding, and also fresh water fisheries. The area of recession cropping was sensitive to the flood volume and shape, with some 100 000–120 000 ha cropped in wet years but only 15 000–50 000 ha during the drought following 1970. The dry season flows were used to support commercially irrigated agriculture, including sugar cane.

The Manantali dam on the Bafing tributary was designed to provide hydroelectric power and irrigation for agriculture, as well as improving navigation. However, the change from recession agriculture to modern irrigation was expected to require a transition period. During this period it was planned to use the reservoir to provide an artificial flood which would provide a more reliable means of inundation than the natural river. The design of the artificial or synthetic flood was based on the shape of the hydrograph of the natural flood, but the flows of the uncontrolled tributaries would be supplemented by releases from the reservoir to give a duration and timing of inundation corresponding to the traditional system of agriculture. This control is illustrated by the flows of the Senegal at Bakel (Fig. 11.12) in some recent years.

DEAD SEA WATER BALANCE

The potential effects of a scheme to generate hydroelectric power by introducing water from the Mediterranean Sea into the Dead Sea were the subject of a review in 1982. The Dead Sea level had been falling steadily because of diversion of Jordan flows upstream, and the difference in elevation between the Mediterranean and the Dead Sea would enable hydroelectric power to be generated from water diverted into the lake, which would also raise the levels of the lake. However, potential changes in water quality introduced an unusual but important factor into the investigation of the water balance of the Dead Sea, as lake evaporation would vary with salinity.

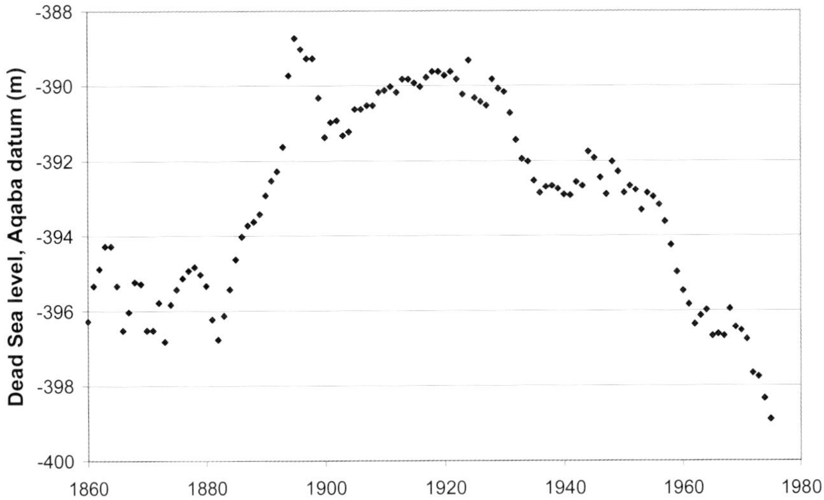

Fig. 11.13 *Dead Sea levels, 1860–1976.*

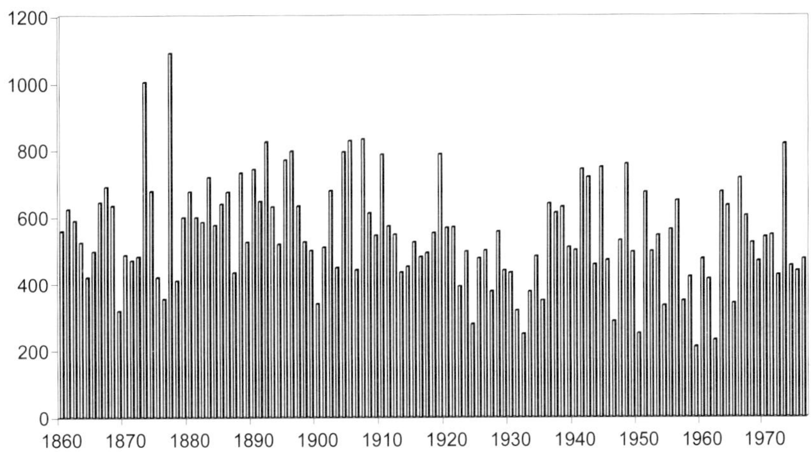

Fig. 11.14 *Jerusalem rainfall (mm), 1860–1976.*

Because the Dead Sea basin is a closed system, with the lake some 400 m below sea level and no outlet except evaporation, the lake levels reflect the balance between inflow and evaporation losses. Annual levels (Fig. 11.13) since 1860 (Klein, 1965; Underhill, 1967) revealed a range of about 10 m over the period of records. After a rise over the period 1880–1900, which must have been due to high rainfall and runoff and may have been linked to the same causes as the high Nile flows during the same period, the level of the Dead Sea has been falling steadily, especially since 1930. This was caused by periods of low rainfall, as recorded at Jerusalem (Fig. 11.14), exacerbated more recently by abstraction of the waters of the Jordan for irrigation.

The main inflow to the Dead Sea is the flow of the River Jordan (area 17 000 km^2), which was measured at the King Hussein Bridge (formerly Allenby Bridge) from 1932

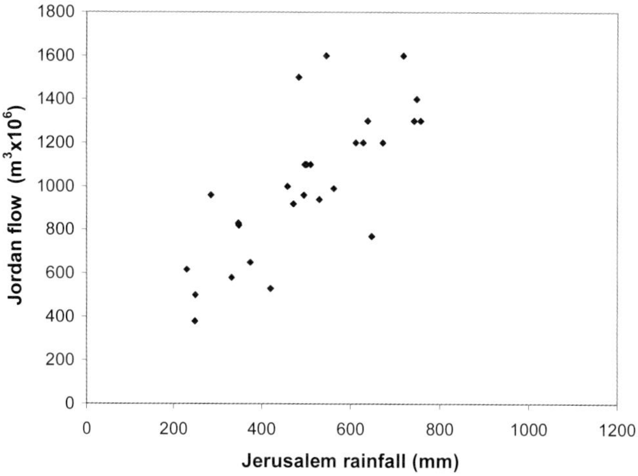

Fig. 11.15 *Jordan flows (at King Hussein Bridge) and Jerusalem rainfall (Old City), 1932–1963.*

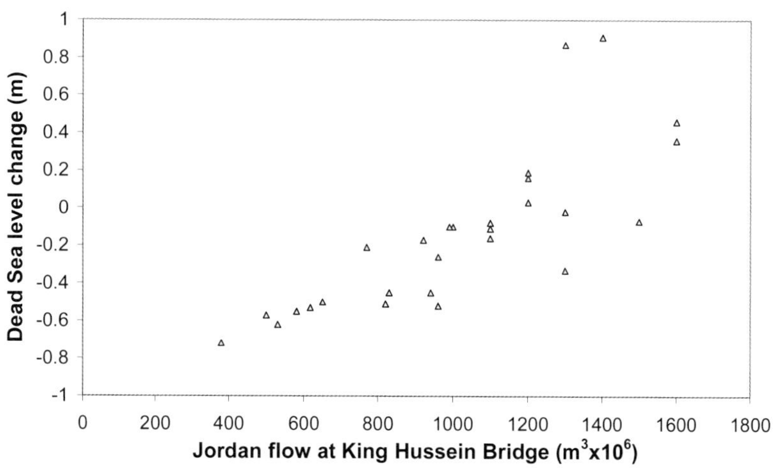

Fig. 11.16 *Jordan flows and Dead Sea level change, 1932–1963.*

to 1963, before irrigation abstractions reached their peak. These flows can be compared (Fig. 11.15) with the long-term rainfall station at Jerusalem to give a reasonable estimate of the long-term mean flow. The rainfall series date back to 1860, and show that recent records come from a relatively dry period; the long-term mean annual rainfall of about 550 mm corresponds to a mean runoff of about 1100 $m^3 \times 10^6$.

The lake balance may be illustrated by a comparison (Fig.11.16) between Jordan inflow and change in lake level, as this river flow forms most of the lake inflow and the evaporation will be more stable than the other components. It is clear that the fall will have been accelerated by a decrease in river inflow following upstream abstractions, which have been estimated to reach eventually 950 $m^3 \times 10^6$ (Weiner & Ben-Zvi, 1982) or most of the Jordan inflow. Alternatively, the lake balance could be illustrated by a comparison between Jerusalem rainfall and change in lake level, during a period when Jordan inflows reflected natural conditions before abstractions were significant.

Over recent years the water abstracted has been balanced by the decrease in lake area and therefore in lake evaporation, and by the stored water released by the lake decline. However, the surface salinity will have increased as the lake fell and the evaporation from the lake surface also decreases with increasing salinity. Evaporation varies with the vapour pressure difference over the water surface interface; the evaporation would almost halve if fresh water were replaced by Dead Sea brine (specific gravity 1.17 g cm^{-3}) at the same temperature, as could be tested by evaporation pans. However, in order to satisfy the energy balance a lake of fresh water would remain at a lower temperature than one of brine. The Penman approach was adapted by Calder & Neal (1984) to give the evaporation rate for saline lakes explicitly in terms of the usual meteorological variables. Whereas the evaporation from the Dead Sea with historic salinity would be 1498 mm with a surface temperature of 26.2°C, the evaporation if the surface water were substituted with Mediterranean Sea water would be 1969 mm at a temperature of 23.6°C.

Because the extraction of potash and other chemicals from the southern Dead Sea is an important contribution to the economies of both Jordan and Israel, the effect of the hydroelectric project on lake level and water quality formed a major component of the potential effects of the project. Alternatively, the rate at which water could be introduced to the lake would be determined by the level at which the Dead Sea level was to be maintained. The water contribution of the project could be equated with the quantity of water abstracted from the River Jordan for irrigation to maintain the long-term equilibrium in lake level, or the sum of the residual Jordan inflow plus the Mediterranean inflow could be compared with evaporation from the lake; this evaporation rate must depend on the resulting surface salinity of the lake. In fact, the Mediterranean-Dead Sea project was shelved, following criticism of its economic feasibility (*Financial Times*, 13 November 1984).

CONCLUSION

These examples illustrate the fact that projects involving changes in the flow or availability of water may have profound effects on the local environment and people. The links between water and environment are so strong that any water engineering project will have repercussions which may not be immediately apparent or straightforward. The example of the Sudd and the Jonglei Canal shows that the chain of effects may be subtle and require detailed research to clarify. The diversity of examples in this chapter hint at the many and wide-ranging effects that changes to the water regime may have on the environment which deserve investigation.

12 POSTSCRIPT

What lessons can be learnt from the examples discussed in this book?

First, time is essential if the hydrologist is to develop a mental picture of the area of study, and the processes which dominate the hydrology of the area. Admittedly, the author has been fortunate to have started his career when time constraints were not as severe as they have become more recently, and in the Nile basin where basic hydrological data had been measured since the beginning of the twentieth century. He was able to spend four years getting to know the wetlands of the upper Nile, which is admittedly a complex area, as a member of a multi-disciplinary team of specialists, and to return to an academic environment to work out the implications of various interactions.

It is unfortunate that less and less time has become available for hydrological investigations, while the basic infrastructure of river flow measurements has become less reliable in many parts of the world. However, the author has made it a rule not to pontificate about the hydrology of an area which he has not seen, though experience has perhaps made it easier to observe the key features in a rapid visit. He believes he shares this habit with his colleagues, though it is not always easy to explain what has been learnt during a visit.

The need may be described briefly as a requirement to observe the hydrological data together with the physical environment, and after talking to local observers to let the combination dictate a narrative account of the hydrology, a procedure which may be best illustrated by reference to some of the examples discussed above.

In several cases—the Tehran area and Varamin plain in Iran, Bahrain island, Oman and Abu Dhabi, and the Betwa basin in central India—it was possible to study areas where the water resources and the traditional water use were, or had been recently, in equilibrium; it was possible to compare the natural surplus—recharge, groundwater flow or runoff—with crop water use.

In the case of the rainfall regime of the Tongariro basin in New Zealand, the fact that the direction of heavy storms had been misinterpreted in drawing up the isohyetal map of the region came to light in discussion with a local geography teacher who happened to be a keen fisherman; this discussion was followed by a visit from the Meteorological Department. In western Iran, the rainfall regime of the Zagros mountain range was explained by a combination of the recorded data and observation of the woodland vegetation. With the Alborz range between Tehran and the Caspian Sea, where both the rainfall regime and the vegetation distribution were more complex, the problem was illuminated by discussion with experienced local meteorologists after a field visit.

The importance of evaporation losses from palm plantations in Bahrain and Oman, and the part this played in the water balance of these areas became clear from tours with colleagues, while the key role of evaporation from the wetlands of the Sudd in the southern Sudan was illuminated by discussion with team colleagues and by a chance meeting with a distinguished Egyptian botanist who had taken part in experimental work nearly 40 years earlier. The role of soil moisture regime in the hydrology of the Betwa basin in central India was made clear in a multi-disciplinary research programme, and this in turn contributed to an archaeological investigation.

The groundwater regime of the Tehran aquifer may have been relatively simple, but its main elements could be observed in reconnaissance with geological colleagues; this led to initial estimates and a recommended measurement programme. The relative simplicity of water balance studies in humid climates has been illustrated by examples from New Zealand, Sarajevo and Patagonia, where the vegetation showed that soil moisture stress was rarely a factor. Although monsoon climates with an extreme contrast between seasons of water surplus and shortage present at first a more complex problem, the deciduous vegetation suggests a simplified approach to the rainfall season as a single storm. The semiarid or arid environment is the most difficult to treat in physical terms and, as in Botswana or Yemen, it may be necessary to treat the area with an empirical model. Water resources projects may be designed to overcome problems of hydrological variability; in addition to seasonal and annual variability it will increasingly be necessary to consider longer-term climate change.

The incidence of floods may be approached either statistically or physically, but the links between flood variability and climate are strong. Environmental factors introduce another dimension to hydrological problems, but a combination of observation, measurement and analysis is as important as in other topics.

REFERENCES

Abbott, M. B. & Refsgaard, J. Ch. (1996) *Distributed Hydrological Modelling*. Water Science and Technology Library 22. Kluwer Academic Publishers, Dordrecht, The Netherlands.

Bader, J. C. (1992) Consignes de gestion du barrage à vocation multiple de Manantali: détermination des côtes limites à respecter dans la retenue. *Hydrol. Continent.* **7**, 3–12.

Bagirathan, V. R. & Shaw, E. M. (1978) Rainfall depth–duration–frequency studies for Sri Lanka. *J. Hydrol.* **37**, 223–239.

Beaumont, P. (1968) Qanats on the Varamin Plain, Iran. *Trans. Inst. Brit. Geogr.* **45**, 169–179.

Beaumont, P. (1971) Qanat systems in Iran. *Hydrol. Sci. Bull.* **16**(1), 39–50.

Bell. J. P. (1976) *Neutron Probe Practice*. Institute of Hydrology Report no. 19, Wallingford, UK.

Bell, J. P. & McCulloch, J. S. G. (1966) Soil moisture estimation by the neutron scattering method in Britain. *J. Hydrol.* **4**, 254–263.

Beran, M. A. & Sutcliffe, J. V. (1972) An index of flood-producing rainfall based on rainfall and soil moisture deficit. *J. Hydrol.* **17**, 229–236.

Bernier, J. (1956) Sur l'application des diverses lois limites des valeurs extrêmes aux débits de crue. *La Houille Blanche* **5**, 718–725.

Beven, K. J. (2001) *Rainfall–Runoff Modelling: The Primer*. John Wiley and Sons, Chichester, UK.

Bishop, W. W. (1969) Pleistocene stratigraphy in Uganda. *Geological Survey Uganda, Memoir 10*. Entebbe, Uganda.

Black, A. R. & Law, F. M. (2004) Development and utilization of a national web-based chronology of hydrological events. *Hydrol. Sci. J.* **49**(2), 237–246.

Blackie, J. R., Edwards, K. A. & Clarke, R. T. (eds) (1979) Hydrological research in East Africa. *E. Afr. Agric. For. J.* **43** (Special Issue).

Bras, R. G. & Rodriguez-Iturbe, I. (1994) *Random Functions and Hydrology*. Dover Books, New York, USA.

Brown, R. H. (1953) Selected procedures for analysing aquifer test data. *J. Am. Water Works Assoc.* **45**, 844–866.

Butzer , K. W. (1971) Recent history of an Ethiopian delta. *University of Chicago Geography Department Research Paper 136*. Chicago, USA.

Calder, I. R. & Neal, C. (1984) Evaporation from saline lakes: a combination approach. *Hydrol. Sci. J.* **29**(1), 89–97.

Centre for Ecology & Hydrology (CEH) (2001) *Regional Flood Frequency Analysis Software (FLOODS)*. CEH, Wallingford, UK.

Chow, V. T. (ed.) (1964) *Handbook of Applied Hydrology*. McGraw-Hill, New York, USA.

Chow, V. T., Maidment, D. R. & Mays, L. W. (1988) *Applied Hydrology*. McGraw-Hill, New York, USA.

Cole, J. A. (1973) Examples of flow-balance solutions for aquifer T and S by an automatic optimisation procedure. *Water Research Association TM 85*. Medmenham, UK.

Cunnane, C. (1978) Unbiased plotting positions—a review. *J. Hydrol.* **37**, 205–222.

Doorenbos, J. & Pruitt, W. U. (1977) *Crop Water Requirements*. Irrigation and Drainage Paper 24. Food and Agriculture Organization, Rome, Italy.

Drew, F. (1875) *The Jummoo and Kashmir Territories: A Geographical Account*. Edward Stanford, London, UK.

Eagleson, P. S. (2002) *Ecohydrology: Darwinian Expression of Vegetation Form and Function*. Cambridge University Press, UK.

El Moushid, B. E. F., El Awad, O. M. A. & Ahmed, S. E. (1997) Environmental effect of the Blue Nile sediment on reservoirs and irrigation canals. In: *5th Nile 2002 Conf.* (Addis Ababa, Ethiopia).

FAO (1993) *A Climatic Database for Irrigation Planning and Management*. Irrigation and Drainage Paper 49. Food and Agriculture Organization, Rome, Italy.

Farquharson, F. A. K. & Sutcliffe, J. V. (1998) Regional variations of African river flows. In: *Water Resources Variability in Africa during the XXth Century* (ed. by E. Servat, D. Hughes, J.-M. Fritsch & M. Hulme) (Proc. Abidjan '98 Conf., Côte d'Ivoire, November 1998), 161–169. IAHS Publ. 252. IAHS Press, Wallingford, UK.

Farquharson, F. A. K., Lowing, M. J. & Sutcliffe, J. V. (1975) Some aspects of design flood estimation. In: *Inspection, Operation and Improvement of Existing Dams* (ed. by A. I. B. Moffat) (Symposium at Newcastle upon Tyne, Sept. 1975), 4.7-1–4.7-9. British National Commission on Large Dams/University of Newcastle upon Tyne, UK.

Farquharson, F. A. K., Meigh, J. R. & Sutcliffe, J. V. (1992) Regional flood frequency analysis in arid and semi-arid areas. *J. Hydrol.* **138**, 487–501.

Farquharson, F. A. K., Sutcliffe, J. V. & Meigh, J. R. (1993) Caractéristiques statistiques de la crue régionale en Afrique de l'Ouest. *Hydrol. Continent.* **8**, 3–16.

Farquharson, F. A. K., Plinston, D. T. & Sutcliffe, J. V. (1996) Rainfall and runoff in Yemen. *Hydrol. Sci. J.* **41**(5), 797–811.

Flood Estimation Handbook (1999) See: Institute of Hydrology (1999).

Flood Studies Report (1975) See: Natural Environment Research Council (1975).

Fréchet, M. (1927) Sur la loi de probabilité de l'écart maximum. *Annales de la Société Polonaise de Mathematiques* **6**, 93–117. Krakow, Poland.

Gardner, C. M. K., Robinson, D. A., Blyth, K. & Cooper, J. D. (2001) Soil water content. In: *Soil and Environmental Analysis: Physical Methods* (second edn) (ed. by K. A. Smith & C. E. Mullins), 1–64. Marcel Dekker Inc., New York, USA.

Garstin, W. (1904) Report upon the basin of the Upper Nile (In a dispatch from His Majesty's Agent and Consul-General at Cairo). *Parliamentary Accounts and Papers 1904* **111**, 315–735.

Gash, J. H. C. (1979) An analytical model of rainfall interception by forests. *Quart. J. Roy. Met. Soc.* **105**, 43–55.

Grijsen, J. G., Snoeker, X. C., Vermeulen, C. J. M., Mohamed el Amin Moh. Nur & Yasir Abbas Mohamed (1992) An information system for flood early warning. In: *Third Int. Conf. on Floods and Flood Management*, 263–289.

Grindley, J. (1970) Estimation and mapping of evaporation. In: *World Water Balance*, vol. 1, 200–213. IAHS Publ. 92. IAHS Press, Wallingford, UK.

Grove, A. T., Street, F. A. & Goudie, A. S. (1975) Former lake levels and climate change in the rift valley of southern Ethiopia. *Geogr. J.* **141**, 177–202.

Gumbel, E. J. (1941) Probability interpretation of the observed return period of floods. *Trans. Am. Geophys. Un.*, Part III, 836–850.

Gunston, H. (1998) *Field Hydrology in Tropical Countries: A Practical Introduction.* Intermediate Technology Publications, London, UK.

Hardy, S., Dugdale, G., Milford, J. R. & Sutcliffe, J. V. (1989) The use of satellite derived rainfall estimates as inputs to flow prediction in the River Senegal. In: *New Directions for Surface Water Modelling* (ed. by M. L. Kavvas) (Proc. Baltimore Symp., May 1989), 23–30. IAHS Publ. 181. IAHS Press, Wallingford, UK.

Herschy, R. (2003) *World Catalogue of Maximum Observed Floods.* IAHS Publ. 284. IAHS Press, Wallingford, UK.

Hewitt, K. (1982) Natural dams and outburst floods of the Karakorum Himalaya. In: *Hydrological Aspects of Alpine and High Mountain Areas* (ed. by J. W. Glen), 259–269. IAHS Publ. 138. IAHS Press, Wallingford, UK.

Hodnett, M. G. & Bell, J. P. (1986) Soil moisture investigations of groundwater recharge through black cotton soils in Madhya Pradesh, India. *Hydrol. Sci. J.* **31**(3), 361–381.

Holder, R. L. (1985) *Multiple Regression in Hydrology.* Institute of Hydrology, Wallingford, UK.

Hosking, J. R. M., Wallis, J. R. & Wood, E. F. (1984) Estimation of the generalised extreme value distribution by the method of probability weighted moments. *Technometrics* **27**, 251–261.

Howell, P. P, Lock, J. M. & Cobb, S. M. (eds) (1988) *The Jonglei Canal: Impact and Opportunity.* Cambridge University Press, UK.

Hurst, H. E. (1952, 1957 2nd enlarged edn) *The Nile.* Constable, London, UK.

Hurst, H. E. & Phillips, P. (1938) *The Hydrology of the Lake Plateau and Bahr el Jebel. The Nile Basin,* vol. V. Government Press, Cairo, Egypt.

Hurst, H. E., Black, R. P. & Simaika, Y. M. (1946) *The Future Conservation of the Nile. The Nile Basin,* vol. VII. Government Press, Cairo, Egypt.

Institute of Hydrology (1978) Nile Waters Study: Hydrology. Report to Sir Alexander Gibb & Partners. Institute of Hydrology, Wallingford, UK.

Institute of Hydrology (1993) *HYDATA v 3.1 Operation Manual.* Institute of Hydrology, Wallingford, UK.

Institute of Hydrology (1999) *Flood Estimation Handbook: Procedures for Flood Frequency Estimation*, five vols. Institute of Hydrology, Wallingford, UK.

Institution of Civil Engineers (1978) *Floods and Reservoir Safety: An Engineering Guide.* Institution of Civil Engineers (ICE), London, UK.

ISO (1993) *Soil Quality—Determination of Dry Matter and Water Content on a Mass Basis—Gravimetric Method.* Doc. ISO 11465. International Standards Organization, Geneva, Switzerland.

ISO (1997a) *Soil Quality—Determination of Dry Matter and Water Content on a Volume Basis—Gravimetric Method.* Doc. ISO 11461. International Standards Organization, Geneva, Switzerland.

ISO (1997b) *Measurement of Liquid Flow in Open Channels—Velocity–Area Methods.* Doc. ISO 748. International Standards Organization, Geneva, Switzerland.

Jenkinson, A. F. (1955) The frequency distribution of the annual maximum (or minimum) values of meteorological elements. *Quart. J. Roy. Met. Soc.* **87**, 158–171.

Jonglei Investigation Team (1954) *The Equatorial Nile Project and its Effects in the Anglo-Egyptian Sudan. Report of the Jonglei Investigation Team*, four vols. Sudan Government, Khartoum.

Kirby, C., Newson, M. D. & Gilman, K. (1991) *Plynlimon Research: The First Two Decades*. Institute of Hydrology, Report no. 109, Wallingford, UK.

Klein, C. (1965) On the fluctuations of the level of the Dead Sea since the beginning of the 19th century. *Hydrological Paper no. 7, Israel Hydrological Service*.

Knight, E. F. (1893) *Where Three Empires Meet*. Longmans Green, London, UK.

Knill, J. L. & Jones, K. S. (1968) Ground-water conditions in Greater Tehran. *Quart. J. Engng Geol.* **1**, 181–194.

Lazenby, J. B. C. & Sutcliffe, J. V. (1994) Transitional control of the River Senegal. In: *Integrated River Basin Development* (ed. by C. Kirby & W. R. White), 503–513. John Wiley & Sons, Chichester, UK.

Linsley, R. K., Kohler, M. A. & Paulhus, J. L. H. (1958) *Hydrology for Engineers*. McGraw-Hill, New York, USA.

Lyons, H. G. (1906) *The Physiography of the River Nile and its Basin*. Survey Department, Cairo, Egypt.

McMahon, T. A. & Mein, R. G. (1978) *Reservoir Capacity and Yield*. Developments in Water Science no. 9. Elsevier, Amsterdam, The Netherlands.

Maidment, D. R. (1993) *Handbook of Hydrology*. McGraw Hill, New York, USA.

Mandeville, A. N., O'Connell, P. E., Sutcliffe, J. V. & Nash, J. E. (1970) River flow forecasting through conceptual models: Part III, The Ray catchment at Grendon Underwood. *J. Hydrol.* **11**, 109–128.

Meigh, J. R. (1995) The impact of small farm reservoirs on urban water supplies in Botswana. *Natural Resources Forum* **19**(1), 71–93.

Meigh, J. R., Farquharson, F. A. K. & Sutcliffe, J. V. (1999) A worldwide comparison of regional flood estimation methods and climate. *Hydrol. Sci. J.* **42**(2), 225–244.

Midgley, D. C. (1952) A preliminary study of the surface water resources of the Republic of South Africa. PhD Thesis, University of Natal, South Africa.

Midgley, D. C. & Pitman, W. V. (1969) *Surface Water Resources of South Africa*. Report no. 2/69. Hydrological Research Unit, University of Witwatersrand, South Africa.

Migahid, A. M. (1948) *Report on a Botanical Excursion to the Sudd Region*. Fouad I University Press, Cairo, Egypt.

Migahid, A. M. (1952) *Further Observations on the Flow and Loss of Water in the "Sudd" Swamps of the Upper Nile*. Fouad I University Press, Cairo, Egypt.

Monteith, J. L. (1985) Evaporation from land surfaces: progress in analysis and prediction since 1948. In: *Advances in Evapotranspiration*, 4–12. Am. Soc. Agric. Engrs.

Moss, M. E. (1982) *Concepts and Techniques in Hydrological Network Design*. WMO Operational Hydrology Report no. 19. World Meteorological Organization, Geneva, Switzerland.

Nash, J. E. (1960) A note on a investigation into two aspects of the relation between rainfall and storm runoff. In: *Surface Waters, Proc. IAHS General Assembly Helsinki*, vol. 1, 567–578. IAHS Publ. 51. IAHS Press, Wallingford, UK.

Nash, J. E. & Sutcliffe, J. V. (1969) Flood-wave formation. In: *Hydrological Forecasting*, 147–155. WMO Technical Note no. 92. World Meteorological Organization, Geneva, Switzerland.

Nash, J. E. & Sutcliffe, J. V. (1970) River flow forecasting through conceptual models: Part I, A discussion of principles. *J. Hydrol.* **10**, 282–290.

Natural Environment Research Council (NERC) (1975) *Flood Studies Report*, five vols. NERC, Wallingford, UK.

Newson, M. D. (1992) *Land, Water and Development: River Basin Systems and Their Sustainable Management*. Routledge, London, UK.

Parks, Y. P. & Gustard, A. (1982) A reservoir storage yield analysis for arid and semi-arid climates. In: *Optimal Allocation of Water Resources* (ed. by M. J. Lowing) (Proc. Exeter Symposium, July 1982), 49–57. IAHS Publ. 135. IAHS Press, Wallingford, UK.

Parks, Y. P. & Smith, P. J. (1983) Factors affecting the flow of aflaj in Oman: a modelling study. *J. Hydrol.* **65**, 293–312.

Parks, Y. P. & Sutcliffe, J. V. (1987) The development of hydrological yield assessment in NE Botswana. In: *Proc. Brit. Hydrol. Soc. National Hydrology Symposium, Hull*, 12.1–12.11.

Penman, H. L. (1948) Natural evaporation from open water, bare soil and grass. *Proc. Roy. Soc.* **193**, 120–145.

Penman, H. L. (1950) The water balance of the Stour catchment area. *J. Instn Water Engrs* **4**(6), 457–469.

Penman, H. L. (1963) *Vegetation and Hydrology*. Commonwealth Agricultural Bureaux, Farnham Royal, UK.

Pereira, H. C. (ed.) (1962) Hydrological effects of changes in land use in some East African catchment areas. *E. Afr. Agric. For. J.* **27** (Special Issue).

Piper, B. S., Plinston, D. T. & Sutcliffe, J. V. (1986) The water balance of Lake Victoria. *Hydrol. Sci. J.* **31**(1), 25–47.

Piper, B. S., Sutcliffe, J. V. & Parks, Y. P. (1994) The Victoria Project, Sri Lanka: Hydrological analysis for Victoria dam. *Proc. Civ. Engrs Water, Marit. & Energy* **106**, 43–52.

Pitman, W. V. (1973) *A Mathematical Model for Generating Monthly River Flows from Rainfall Data in South Africa.* Report no. 2/73. Hydrological Research Unit, University of Witwatersrand, South Africa.

Price, M. (1996) *Introducing Groundwater*. Nelson Thornes, London, UK.

Reed, D. W. (1999) *Flood Estimation Handbook* vol. 1: *Overview*. Institute of Hydrology, Wallingford, UK.

Rodda, J. C. (1962) An objective method for the assessment of areal rainfall amounts. *Weather* **17**, 54–59.

Rodier, J. A. & Roche, M. (1984) *World Catalogue of Maximum Observed Floods*. IAHS Publ. 143. IAHS Press, Wallingford, UK.

Rosenbrock, H. H. (1960) An automatic method of finding the greatest or least value of a function. *Computer J.* **3**.

Rushton, K. R. (2003) *Groundwater Hydrology: Conceptual and Computational Models*. John Wiley & Sons, Chichester, UK.

Sene, K. J. & Plinston, D. T. (1994) A review and update of the water balance of Lake Victoria in East Africa. *Hydrol. Sci. J.* **39**(1), 47–63.

Shahin, M. (2002) *Hydrology and Water Resources of Africa*. Kluwer Academic Publishers, Dordrecht, The Netherlands.

Shaw, E. M. (1994) *Hydrology in Practice*, third edn. Chapman Hall, London, UK.

Shaw, J. (2000) Sanchi and its archaeological landscape: Buddhist monasteries, settlements and irrigation works in central India. *Antiquity* **74**, 775–776.

Shaw, J. & Sutcliffe, J. V. (2001) Ancient irrigation works in the Sanchi area: an archaeological and hydrological investigation. *South Asian Studies* **17**, 55–75.

Shaw, J. & Sutcliffe, J. V. (2003) Ancient dams, settlement archaeology and Buddhist propagation in central India: the hydrological background. *Hydrol. Sci. J.* **48**(2), 277–291.

Smith, P. J. & Piper, B. S. (1978) A non-linear optimisation method for the estimation of aquifer parameters. *J. Hydrol.* **39**, 255–271.

Strangeways, I. C. (2003) *Measuring the Natural Environment* (second edn). Cambridge University Press, UK.

Sumanasekere, P. (1977) An analysis of rainstorms in the Upper Mahaweli catchment. Meteorological Department, Colombo, Sri Lanka. Unpublished report.

Sutcliffe, J. V. (1957) The hydrology of the Sudd region of the upper Nile. PhD Thesis, University of Cambridge, UK.

Sutcliffe, J. V. (1966) The assessment of random errors in areal rainfall estimation. *Hydrol. Sci. Bull.* **11**(3), 35–42.

Sutcliffe, J. V. (1967) Report on ground-water extraction on Bahrain Island and coastal area. British Geological Survey/Institute of Hydrology Report to ODA.

Sutcliffe, J. V. (1974) A hydrological study of the southern Sudd region of the upper Nile. *Hydrol. Sci. Bull.* **19**(2), 237–255.

Sutcliffe, J. V. (1975) Choice of estimation techniques. In: *Flood Studies Conference*, 67–69. Institution of Civil Engineers, London, UK.

Sutcliffe, J. V. (1978) *Methods of Flood Estimation: a Guide to the Flood Studies Report*. Report no. 49, Institute of Hydrology, Wallingford, UK.

Sutcliffe, J. V. (1987) The use of historic records in flood frequency analysis. *J. Hydrol.* **96**, 159–171.

Sutcliffe, J. V. (1990) The use of hydrological models in water resources planning. In: *United Nations International Workshop on Water Resources Planning and Management in China* (ed. by R. M. North), 107–133. UNDTCD/SSTC, New York, USA.

Sutcliffe, J. V. (1995) Hydrology in the developing world. In: *Time and the River: Essays by Eminent Hydrologists* (ed. by G. W. Kite), 59–92. Water Resources Publications, LLC, Highlands Ranch, Colorado, USA.

Sutcliffe, J. V. (1996) The role of local hydrological studies in the water resources management of the Nile basin. In: *4th Nile 2002 Conference* (Kampala, Uganda).

Sutcliffe, J. V. & Carpenter, T. G. (1967) The assessment of runoff from a mountainous and semi-arid area in western Iran. In: *Hydrological Aspects of the Utilization of Water* (IAHS General Assembly of Bern), 383–394. IAHS Publ. 76, IAHS Press, Wallingford, UK.

Sutcliffe, J. V. & Farquharson, F. A. K. (1998) Flood frequency studies using regional methods. In: *Statistical and Bayesian Methods in Hydrological Sciences* (ed. by E. Parent, P. Hubert, B. Bobée & J. Miquel), 341–355. IHP-V Technical Documents in Hydrology, UNESCO, Paris, France.

Sutcliffe, J. V. & Green, C. S. (1986) Water balance investigations of recharge in Madhya Pradesh, India. *Hydrol. Sci. J.* **31**(3), 383–394.

Sutcliffe, J. V. & Knott, D. G.(1987) Historical variations in African water resources. In: *The Influence of Climate Change and Climatic Variability on the Hydrologic Regime and Water Resources* (ed. by S. I. Solomon, M. Beran & W. Hogg) (Proc. Vancouver Symp., August 1987), 463–475. IAHS Publ. 168, IAHS Press, Wallingford, UK.

Sutcliffe, J. V. & Lazenby, J. B. C. (1989) Hydrology and river control on the Niger and Senegal. In: *The Sahel Forum* (Ouagadougou, Burkina Faso), 611–620. International Water Resources Association, Urbana, Illinois, USA.

Sutcliffe, J. V. & Lazenby, J. B. C. (1990) Hydrological data for surface water resources assessment. In: *The Hydrological Basis for Water Resources Management* (ed. by U. Shamir & Chen Jiaqi) (Proc. Beijing Symp., October 1990), 251–259. IAHS Publ. 197, IAHS Press, Wallingford, UK.

Sutcliffe, J. V. & Parks, Y. P. (1982) *A Hydrological Estimate of the Effects of the Jonglei Canal on Areas of Flooding*. Report to FAO. Institute of Hydrology, Wallingford, UK.

Sutcliffe, J. V. & Parks, Y. P. (1987) Hydrological modelling of the Sudd and Jonglei Canal. *Hydrol. Sci. J.* **32**(2), 143–159.

Sutcliffe, J. V. & Parks, Y. P. (1989) Comparative water balances of selected African wetlands. *Hydrol. Sci. J.* **34**(1), 49–62.

Sutcliffe, J. V. & Parks, Y. P. (1996) Hydrological controls of Sudd ecology. In: *Water Management and Wetlands in Sub-Saharan Africa* (ed. by M. C. Acreman & G. E. Hollis), 51–72. IUCN, Gland, Switzerland.

Sutcliffe, J. V. & Parks, Y. P. (1999) *The Hydrology of the Nile*. IAHS Special Publ. 5. IAHS Press, Wallingford, UK.

Sutcliffe, J. V. & Piper, B. S. (1986) Bilan hydrologique en Guinée et Togo-Bénin. *Hydrol. Continent.* **1**, 51–61.

Sutcliffe, J. V. & Rangeley, W. R. (1960a) Variability of annual river flow related to rainfall records. In: *IAHS Commission of Surface Waters (Helsinki Assembly)*, 182–192. IAHS Publ. 51, IAHS Press, Wallingford, UK.

Sutcliffe, J. V. & Rangeley, W. R. (1960b) An estimation of the long-term yield of a large aquifer at Tehran. In: *IAHS Commission of Subterranean Waters (Helsinki Assembly)*, 264–271. IAHS Publ. 52, IAHS Press, Wallingford, UK.

Sutcliffe, J. V. & Swan, C. H. (1970) The prediction of actual evaporation in semi-arid areas. In: *Symposium on World Water Balance*, 213–224. IAHS Publ. 92, IAHS Press, Wallingford, UK.

Sutcliffe, J. V. & Widgery, N. J. (1997) The problems of sustainable water resources management in Sudan. In: *Sustainability of Water Resources under Increasing Uncertainty* (ed. by D. Rosbjerg, N.-E. Boutayeh, A. Gustard, Z. W. Kundzewicz & P. F. Rasmussen) (Proc. Rabat Symp., April 1997), 259–265. IAHS Publ. 240, IAHS Press, Wallingford, UK.

Sutcliffe, J. V., Agrawal, R. P. & Tucker, J. M. (1981) The water balance of the Betwa basin, India. *Hydrol. Sci. Bull.* **26**(2), 149–158.

Sutcliffe, J. V., Dugdale, G. & Milford, J. R. (1989) The Sudan floods of 1988. *Hydrol. Sci. J.* **31**(3), 355–364.

Tate, E. L., Sene, K. J. & Sutcliffe, J. V. (2001) A water balance study of the upper White Nile basin flows in the late nineteenth century. *Hydrol. Sci. J.* **46**(2), 301–318.

Tate, E. L., Sutcliffe, J. V., Conway, D. & Farquharson, F. A. K. (2004) Water balance of Lake Victoria: update to 2000 and climate change modelling to 2100. *Hydrol. Sci. J.* **49**(4), 563–574.

Thornthwaite, C. W. (1948) An approach toward a rational classification of climate. *Geogr. Rev.* **38**, 85–94.

Turc, L. (1954) Calcul du bilan de l'eau: évaluation en function des précipitations et des températures. In: *Rapports de la Commission des Eaux de Surface*, vol. III (IAHS Rome Symp.), 188–202. IAHS Publ. 38, IAHS Press, Wallingford, UK.

Underhill, H. W. (1967) Dead Sea levels and the PEF mark. *Palestine Exploration Quarterly*.

Vieux, B. E. (2001) *Distributed Hydrologic Modeling Using GIS*. Water Sciences and Technology Library 38. Kluwer Academic Publishers, Dordrecht, The Netherlands.

Wadia, D. N. (1975) *Geology of India*. Tata McGraw-Hill, New Delhi, India.

Ward, R. C. & Robinson, M. (2000) *Principles of Hydrology* (fourth edn). McGraw Hill, London, UK.

Watson, K. K. (1966) An instantaneous profile method for determining the hydraulic conductivity of porous materials. *Water Resour. Res.* **2**(4), 709–715.

Weiner, D. & Ben-Zvi, A. (1982) A stochastic dynamic programming model for the operation of the Mediterranean-Dead Sea Project. *Water Resour. Res.* **18**(4), 729–734.

Wenzel, L. K. (1942) Methods for determining permeability of water-bearing materials. *US Geological Survey, Water-Supply Paper no. 887.*

Willis, R. P. (1967) Geology of the Arabian Peninsula: Bahrain. *US Geological Survey Prof. Paper 560-E.*

Wilson, E. M. (1990) *Engineering Hydrology* (fourth edn). Macmillan, London, UK.

Wiltshire, S. E. (1986) Regional flood frequency analysis. II: Multivariate classification of drainage basins in Britain. *Hydrol. Sci. J.* **31**(3), 335–346.

World Meteorological Organization (1994) *Guide to Hydrological Practices*. WMO, Geneva, Switzerland.

INDEX

The Basis of Civilization – Water Science?

Edited by John C. Rodda & Lucio Ubertini

IAHS Publ. 286 (*July 2004*) ISBN 1-901502-57-0
342 + x pp. Price £58.50

- *Can history teach us how to combine politics and religion with water science and technology?*
- *What is the current role of water science?*

Trevi fountain, Rome by Marco Mercuri, reproduced with permission.

Selected edited papers from the international symposium held in Rome (December 2003), which address the theme from different angles to provide a fascinating collection that will broaden the perspectives of all involved in hydrology.

In his introduction Jim Dooge explores the linkages between hydrology and society throughout history and outlines the slow development of water technology and the even slower development of water science. Other authors provide examples of the **rise of the understanding of water science.** The **impact of water resources development and management on society** is discussed from political, economic and cultural viewpoints. The **approaches to risk and conflict** involving water are considered in detail and issues of **governance, past and present**, are also reviewed. Contributions from the UN (the achievementsof the UN International Year of Freshwater, 2003); FAO (water and food security), and IAEA (the on-going contribution to isotope hydrology), are included.

Scales in Hydrology and Water Management

Echelles en hydrologie et gestion de l'eau

Edited by Ioulia Tchiguirinskaia,
Mike Bonell & Pierre Hubert

A joint IAHS / UNESCO-IHP publication

A collection of essays by engineers and scientists currently tackling the issues of scale dependence in hydrological phenomena and water resources management

Space and time scales are fundamental for both scientific and operational hydrology issues. The question of how different scales interact and how we can transport results from one scale to another is basic to hydrological understanding and strategically important for decision-making.

Two questions dominate water resources management: how much remains for the future? and what is the response to environmental stresses? To answer these a proper grasp of the interactions at different scales between the atmosphere, lithosphere, hydrosphere and biosphere, and their relations to human activities is needed.

IAHS Publication 287 (*September 2004*)
ISBN 1-901502-47-3; 170 + x pp.; Price £36.95

Floods recur over time scales of years but can impact in minutes

Climate changes at the macroscale (10^6 m), but at the microscale (10^{-2} m) hydrology is linked though fluid mechanics and hydraulics to the physical sciences

IAHS AISH

www.iahs.info

INTERNATIONAL ASSOCIATION OF HYDROLOGICAL SCIENCES

IAHS, established in 1922, is a nonprofit-making nongovernmental scientific organization committed to serving the science of hydrology and the worldwide community of hydrologists

IAHS is concerned with all components of the hydrological cycle and related processes, as well as the application of hydrology and its implications for society and the environment. Currently nine International Commissions deal with the hydrological cycle, water for society and the environment, and specific techniques:

- *Surface Water*
- *Groundwater*
- *Continental Erosion*
- *Snow and Ice*
- *Water Quality*
- *Water Resources Systems*
- *Coupled Land–Atmosphere Systems*
- *Remote Sensing*
- *Tracers*

A recent (2003) initiative on Prediction in Ungauged basins (PUB) involves all the Commissions.

***Information** about IAHS, the Commissions, PUB, and their activities is available at* www.iahs.info, *or from the IAHS Secretary General:*

Dr Pierre Hubert, Secretary General IAHS,
Ecole des Mines de Paris, 35 Rue Saint-Honoré, F-77305 Fontainebleau, France
e-mail: *iahs@ensmp.fr*

Individual membership of IAHS is *free of charge* and open to anyone wishing to participate in IAHS activities, such as IAHS symposia and workshops; worldwide there are now more than 4000 members. The registration form can be downloaded from the IAHS website or obtained from:

Mrs Jill Gash, Membership Secretary,
IAHS Press, Centre for Ecology and Hydrology, Wallingford, Oxfordshire OX10 8BB, UK
e-mail: *jilly@iahs.demon.co.uk*

Members receive the Newsletter and information about IAHS activities, and can obtain discounts on Hydrological Sciences Journal and most other IAHS publications; at least 25% discount to all members, and 80% discount for members in financially disadvantaged countries.

IAHS Publications IAHS has a long tradition of publishing titles on topics in hydrology and on related themes. The IAHS Red Books (ISSN 0144-7815), provides a means for international dissemination of research papers from authors worldwide. Many of the publications are collections of peer-reviewed papers selected from the proceedings of IAHS-convened and other international meetings. Others are reports on specific topics. IAHS also publishes *Hydrological Sciences Journal*.. Further information at *www.iahs.info*, or ask *jilly@iahs.demon.co.uk* (address as above) for catalogues and flyers.

Distribution of IAHS Publications to "Countries in Need" is funded by the IAHS Task Force for Developing Countries (TFDC). *Hydrological Sciences Journal* and other IAHS publications are distributed free to more than 70 organizations worldwide, to help scientists there to acquire pioneering results of hydrological sciences and their application. The publications are provided on the understanding that they are made available to specialists both inside and outside the recipient organizations. All correspondence concerning the distribution of IAHS publications to "countries in need" should be sent to:

Prof. Xia Jun, Department of Hydrology and Water Resources, College of Water Resources & Hydropower Engineering, Wuhan University, East Lake Road no. 8, Wuhan 430073, China
e-mail: *xiaj@igsnrr.ac.cn*